CO-ORDINATION
OF
GALACTIC RESEARCH

INTERNATIONAL ASTRONOMICAL UNION
SYMPOSIUM No. 7
HELD AT SALTSJÖBADEN NEAR STOCKHOLM, JUNE 1957

SECOND CONFERENCE ON CO-ORDINATION OF GALACTIC RESEARCH

EDITED BY

A. BLAAUW, G. LARSSON-LEANDER, N. G. ROMAN, A. SANDAGE, H. F. WEAVER AND A. D. THACKERAY

Printed with financial assistance from U.N.E.S.C.O.

CAMBRIDGE
AT THE UNIVERSITY PRESS
1959

CAMBRIDGE
UNIVERSITY PRESS

University Printing House, Cambridge CB2 8BS, United Kingdom

Cambridge University Press is part of the University of Cambridge.

It furthers the University's mission by disseminating knowledge in the pursuit of education, learning and research at the highest international levels of excellence.

www.cambridge.org
Information on this title: www.cambridge.org/9781316612620

First published 1959
First paperback edition 2016

A catalogue record for this publication is available from the British Library

ISBN 978-1-316-61262-0 Paperback

CONTENTS

PREFACE

The editorial work in preparing the present report has been shared by six of the participants of the conference. Section I was edited by Dr A. Sandage, sections II A to II E by Dr G. Larsson-Leander, section II F by Dr H. F. Weaver, section II G by Dr Nancy G. Roman and section III by Dr A. D. Thackeray. The remaining parts and the co-ordination of the various sections were done by the undersigned.

In accordance with a decision taken at the Stockholm conference, an attempt has been made to present the material in the same way as the report of the Groningen Conference of 1953 (see also the Introduction). Differences have remained between the conceptions of the work of the editors with regard to the amount of detail represented, the frequency of including references to the literature, and the completeness with which the names of the participants in the discussions are mentioned. Little effort has been made to smooth these differences which, I believe, are not disturbing and add a certain variety in the way of presentation of the subjects. I should like to express my indebtedness to my co-editors for their pleasant co-operation in preparing this volume.

A. BLAAUW

GRONINGEN
KAPTEYN LABORATORY
February 1958

INTRODUCTION

Organization and character of the conference

The Second Conference on Co-ordination of Galactic Research was held from 17 to 22 June 1957 at Saltsjöbaden near Stockholm, at the invitation of Dr B. Lindblad, director of the Stockholm Observatory. It was prepared by the sub-commission on Co-ordination of Galactic Research of Commission 33 of the International Astronomical Union. The local organization was in the hands of the Stockholm Observatory, particularly of Professor B. Lindblad and Dr J. Ramberg. By a grant from the Swedish Government the participants received free board and lodging at Vår Gård, a boarding-school belonging to the Swedish Co-operative Union. A considerable part of the travel expenses was defrayed from funds made available by UNESCO and the International Astronomical Union.

The aim of the conference was the same as that of the 1953 Groningen Conference on Co-ordination of Galactic Research, which has been reported on in *Symposium No. 1 of the International Astronomical Union* (Cambridge University Press, 1955). However, whereas the Groningen Conference served as a general survey of problems deserving the attention of workers in galactic structure, the Stockholm discussions were of a more specific nature, dealing with problems whose importance had become particularly evident in the course of the past four years. The first session was devoted to work on star clusters, the investigations of which have become so intimately related to those of galactic structure. The important role of investigations of the Magellanic Clouds in connexion with galactic research was stressed by a special session on the Clouds following the sessions of galactic problems. The various topics were arranged approximately in the order of the Groningen Report.

Most of the subjects were introduced by one of the participants with a review of recent work, and these introductions were followed by discussions of varying length. In accordance with a decision taken at the Conference the present report has the same character as that of the Groningen Conference in that it does not give the introductory papers in full, but rather describes their main points and relevant parts of the discussions. It was considered that such a report, while lacking in the precision with which the introductory remarks are reproduced, has the advantage of conciseness. Since the emphasis of the meeting was on the discussions rather than on the introductory papers, we shall not list all the speakers but refer to the report itself for their names.

Participants

Participants from outside Sweden were: W. Baade, A. Blaauw, D. Chalonge, O. J. Eggen, Ch. Fehrenbach, W. Fricke, H. Haffner, G. Haro, O. Heckmann, D. S. Heeschen, A. R. Hogg, F. J. Kerr, E. K. Kharadze, B. V. Kukarkin, P. G. Kulikovsky, W. W. Morgan, J. J. Nassau, J. H. Oort, P. Th. Oosterhoff, P. P. Parenago, L. Perek, N. G. Roman, A. Sandage, A. D. Thackeray, A. N. Vyssotsky, M. F. Walker, H. F. Weaver, G. Westerhout, R. v. d. R. Woolley; whereas the following Swedish astronomers participated: A. Elvius, T. Elvius, E. Holmberg, G. Larsson-Leander, B. Lindblad, P. O. Lindblad, K. Lodén, L. O. Lodén, K. Lundmark, E. Lyttkens, K. G. Malmquist, J. M. Ramberg, A. Reiz, O. Rydbéck, C. Schalén, U. Sinnerstad, I. Torgård, Å. Wallenquist.

Conclusions reached at the conference

The principal conclusions of the conference are summarized in section IV, which is largely based on the summarizing talk by J. H. Oort. This section includes mentioning of a small committee, appointed in order to explore possible ways of investigating the density distribution of various population components in the region around the sun.

A proposal for establishing photometric standards to be used in cluster work is added as an appendix to this report. It was drawn up by a small committee during the conference, consisting of H. Haffner, P. Th. Oosterhoff, A. Sandage and M. F. Walker (chairman).

I. CLUSTERS AND ASSOCIATIONS

(A) INTRODUCTION

Renewed interest in the study of star clusters and associations has been aroused by at least three developments.

(a) Interpretation of the color-magnitude (C-M) diagrams of clusters and associations with a theory of stellar evolution promises to give data on galactic structure problems. Information of kinematic relations, origins, and past history of stars over the face of the H-R diagram will be available when proper evolutionary interpretation is achieved. Only by assembling C-M diagrams of many clusters of different ages and richness can the empirical approach to evolution be fully exploited.

(b) Clusters and associations are tracers of spiral structure. Modern multi-color photometric methods, developed principally by Becker and by Johnson and Morgan, permit precise determination of distances of clusters. Both Kerr and Oort emphasized the importance of absolute distances in order to calibrate the distance scale of the radio galaxy. If enough clusters and associations are available to define the spiral structure, then the arm spacing and the value of the A constant can be found. This requires accurate, standardized photometry and radial velocity measurements on many clusters.

(c) C-M diagrams for clusters provide the most accurate method for finding the absolute magnitudes of luminous O and B stars. When diagrams for OB clusters are determined as faint as the A and F stars, calibration of the OB stars is immediate. When enough cluster diagrams are available, the cosmic dispersion σ about the mean absolute magnitude M_o will be found. Both M_o and σ are required for tracing spiral arms by O and B stars in associations.

(B) CLUSTERS AND STELLAR EVOLUTION

Precision C-M diagrams for a few nearby clusters are available principally through the photo-electric work of O. J. Eggen and H. L. Johnson. A composite C-M diagram[1] for all the available clusters shows the following systematic regularities: (a) The bright end of the main sequence in different clusters terminates at different absolute magnitudes. (b) The slope of the main sequence near the termination points is always steeper than the corresponding slope for clusters with brighter termination points. (c) When giants or super-giants are present, the absolute magnitude of the giant or super-giant sequence is nearly the same as that of the main sequence termination point. (d) A Hertzsprung gap is present in all clusters whose main sequence termination points are brighter than visual absolute magnitude $M_v = +2 \cdot 5$ but the width of the gap depends on absolute magnitude. These results were known and discussed by Trumpler and have been emphasized by modern work. One interpretation follows from work on stellar evolution by M. Schwarzschild and his school. This group believes that stars with a range of masses contract from the inter-stellar medium and populate the main sequence over a range of luminosities. After thermonuclear processes convert about 10 % of the H to He in any main sequence star, that star leaves the main sequence, moves rapidly to the right in the H-R diagram on a very short time scale (form-ing the Hertzsprung gap), then slows down to form the giant and super-giant sequences. The absolute magnitude of the termination point of the main sequence changes with time in any cluster because the massive, luminous stars burn 10 % of their hydrogen and leave the main sequence before the less luminous stars have evolved very far.

Application of this theory gives the ages of any cluster whose main sequence termination luminosity is known. Ages range from 10^6 years for h and χ Persei to about 5×10^9 years for clusters like M 67. An interesting consequence of these ideas is the large cosmic dispersion expected in the absolute magnitudes of Morgan and Keenan's luminosity classes. In any given cluster, the dispersion will be small or zero. But, because of the change of slope near the main sequence termination point (item b in the listing of the systematic features) stars of the same spectral class near this point will have different absolute magnitudes in clusters of slightly dif-ferent age. This must be borne in mind in calibration of the Morgan-Keenan luminosity classes for stars in the general field. The same is true for the giant and super-giant stars. Generalization of the composite cluster diagram suggests that a continuous variation of absolute magnitude exists for non-main sequence stars from $M_v = +2$ to $M_v = -6$. Because the MK classi-fication recognizes only the five major non-main sequence classes IV, III, II, I b and I a, there will be appreciable cosmic scatter about M_0.

(c) A STANDARDIZED CLUSTER PROGRAM

Precision C-M diagrams for *many* clusters of all ages are needed to guide theoretical workers in stellar evolution, to provide an optical scale for the radio galaxy, and to solve the absolute magnitude calibration problem. The conference participants stressed the extreme importance for all workers in the cluster field to adopt standardized techniques and procedures to insure that photometric data collected in the next few years will be precise and homogeneous. Only in this way can individual results be combined and discussed as a whole. The conference addressed itself to this problem of procedure and discussed two separate sub-problems: (*a*) photometry, proper motion and radial velocity work in clusters already known, and (*b*) discovery of new and more distant clusters. Haffner, in his introductory talk, stressed the lack of detailed knowledge of even the Trumpler class for a large number of known galactic clusters. This is illustrated by means of Table 1, the upper part of which shows the numbers of clusters

Table 1. *Statistics of photometry and spectroscopic classifications available for clusters in Trumpler's catalog*

	North of −15° decl.		South of −15° decl.	
Trumpler type	C-M diagram available	C-M diagram not available	C-M diagram available	C-M diagram not available
0	2	4	2	2
b	15	5	1	19
b–a	6	6	0	8
a	11	9	2	6
f	2	0	0	0
All types	36	24	5	35
Type unknown	25	84	9	117

with known Trumpler type, with and without known C-M diagrams, whereas the bottom line gives the corresponding numbers for the clusters without Trumpler type in Trumpler's catalog. The subdivision according to the northern and southern declinations shows the large deficiency of C-M observations of southern clusters. Haffner urged (*a*) that a co-operative program be undertaken of multi-color photometry for C-M diagrams of most known galactic clusters, and (*b*) that a search program for the discovery of distant clusters be started. The result of such a program would be primarily the optical location and distance calibration of the spiral arms. The time for completion of such a program would be five years. He urged uniformity of photometric technique and suggested adoption of the Johnson U-B-V photometric system and the Johnson-Morgan technique of absorption determination. Additional wave-lengths could be added for special purposes but at least U, B, and V values should be published for each cluster.

The experience of the Hamburg observers with Schmidt plates and modern iris diaphragm photometers shows that high *internal* precision can be obtained for magnitude measures using photographic techniques. The Mount Wilson observers have also had success with large reflector plates measured with a variable iris photometer. The most efficient way to complete the large cluster program therefore seems to be to set up accurate photo-electric scales and zero points near each cluster and then to use photographic interpolation for those clusters with many stars. The conference emphasized the necessity that absolute calibration be done *photo-electrically*. Otherwise the program need not be started because the necessary precision cannot be obtained by photographic transfer methods. The conference discussed at length the needs and procedures for the photo-electric scale and zero-point determinations. A small committee consisting of Haffner, Oosterhoff, Sandage, and Walker (chairman) was appointed to work out details. Their recommendations were approved by the conference and appear in detail in the Appendix. A summary of the conclusions follows:

(*a*) Adopt the U-B-V photometric system of Johnson as the standard to which to refer all instrumental magnitudes.

(*b*) Set up zero-point and photometric scales by photo-electric means in or near each cluster to be studied.

(*c*) For ease of conversion of instrumental magnitudes to the U-B-V system, Walker and Hardie will extend the list of U-B-V standards evenly around the sky in two zones, one at declination $+45°$ and the other at either declination $0°$ or $-10°$. The $45°$ zone will contain six regions separated by 4 hours of R.A. The $0°$ zone will contain eight regions separated by 3 hours of the R.A. Each region will contain four stars; one red and one blue star of 8th magnitude, and one red and one blue star of 10th magnitude. These stars will serve as zero-point stars and as stars to determine the transformation equations from the observer's natural photo-electric system to the U-B-V system.

(*d*) For rich clusters where complete photo-electric coverage is impractical, photographic plates will be used for interpolation between the photo-electric standards. Modern iris photometers must be used for the measurements of the plates.

The Hamburg Observatory has offered to co-operate with any group interested in this cluster program. They will furnish photographic plates in three colors taken with their Schmidt telescope, if any other group will furnish photo-electric standards.

(D) TRUMPLER'S CATALOG

Weaver reported that Trumpler was preparing his extensive cluster data for publication at the time of his death. The material consists of (a) a catalog which lists the X and Y co-ordinates, the radial velocity, the spectral type on Trumpler's own system (which is close to the Yerkes MK-system) and estimates of colors and magnitudes for nearly 5000 stars in 100 clusters; and (b) identification charts for each cluster with the X and Y co-ordinates superposed and each star of the catalog marked. At the time of his death, Trumpler was attempting a conversion of his photometry to the B, V system. This was only partially completed. Weaver stated that the identification charts and the radial velocities were, however, completed and were nearly ready for the printer. The conference urged Weaver to continue with the preparation of Trumpler's material for the press. Both Weaver and the conference delegates stated that Trumpler's color and magnitude estimates should be published to guide future photometric observers, but that no large effort should be made to put these values on the B, V system because present-day requirements will necessitate photometric re-observation in any case. Time is the important point with regard to Trumpler's data. The material will be of such great value in guiding the precision photometric job on many clusters that it should be published as soon as possible. Weaver estimated that the entire material would be in the hands of the printer by the fall of 1958.

(E) DISCOVERY OF NEW CLUSTERS

The discovery and study of very distant clusters is important for the calibration of the galactic distance scale adopted in the interpretation of radio observations. But complete discovery of the nearby loose clusters is also important for studies of the age distribution of clusters because this may give information on star formation rates as a function of time. Two separate discovery techniques are available. These are: (a) inspection of direct photographs to identify clumpings of stars; (b) inspection of objective-prism plates taken with Schmidt cameras to identify a spectral type-apparent magnitude relation.

Haffner's experience with method (a) shows that loose, large diameter aggregates are easier to discover with instruments of small focal length and small aperture than with large Schmidt cameras. For example, Haffner reported discovery of twenty possible new clusters in Canis Major and Puppis with a 120-cm focal length telescope in South Africa. These

7

groupings can be located only with difficulty on the Palomar Sky Atlas. The many faint background stars decrease the contrast between the cluster and the field and the cluster is lost on plates reaching to faint magnitudes.

Nassau reported on Stock's work on the location of new clusters by the objective-prism technique. Stock used a 4° prism on the Cleveland Schmidt and found twenty-one suspected clusters by this method, each of which showed a similarity of spectral type at a given apparent magnitude. The suspected clusters were found from survey plates taken along the galactic plane in a band ±6° wide in latitude and extending from longitude 200° through 0° to 340°. The limit of the survey was about 10·2 photographic magnitude. Three natural groups emerged from Stock's data. These were B clusters, B-A clusters, and F clusters where the F stars were probably giants.

The Cleveland group plans to extend the program to fainter magnitudes to check for the required H-R diagram which must be present for a real cluster. The new survey will again extend from $l = 200°$ through 0° to 340° in a belt ±6° wide, and will reach 13·2 photographic magnitude with the 2° prism. The plate material will be so extensive that Nassau proposes to make copies of the plates available to interested observatories at the cost of reproduction. Tests of the quality of reproduction made by the Eastman Kodak Company showed that spectral classification was as precise on copies as on the original plates themselves. The discovery plan would then be to search these plates for spectral groupings with the expected apparent magnitude range necessary for a real cluster. In this way a large number of clusters which are too loose to be easily picked up from direct plates would be discovered. Nassau estimated that about 370 plates would cover the galactic belt available from Cleveland. The estimated cost of reproduction of 370 plates is about $300.

Lindblad mentioned that the Stockholm Observatory has started a similar discovery survey for southern clusters using the ADH Schmidt of the Boyden Station. There will be some overlap with the Cleveland survey so that a systematic comparison can be made of completeness factors. The dispersion of the Stockholm survey will be nearly the same as that used by the Cleveland group.

Haro stated that the Tonantzintla objective-prism survey covers a belt 10° wide centered on the galactic plane extending from the Coal Sack in the south to Cassiopeia in the north to a limiting magnitude of 13. Copies of these plates can be made available on the same basis as the Cleveland survey at cost to any observatory wishing systematically to work up the material.

Weaver mentioned that the certainty that a given spectral grouping was really a physical cluster was difficult, especially when the grouping is loose and submerged in a dense background field. Nassau agreed and stated that the grouping merely provided a finding list from which objects could be chosen for extensive study. All agreed that to prove existence of a physical grouping requires proper motion study, accurate photometry, and some radial velocity work. Weaver emphasized the importance of the spectral survey method of discovery because here, for the first time, was a way to pick up a class of clusters which have relatively few stars, which often begins at a late spectral class (A or F), and which may be the most numerous class of clusters in the sky but which we know next to nothing about. A nearby example of such a cluster is Coma Berenices which Trumpler showed[2] to have less than fifty stars brighter than visual absolute magnitude +6. Such a cluster would be undiscovered by inspection of direct plates if it were at a greater distance because it would merge into the background field.

Heckmann illustrated not only the fruitfulness of detailed study of such loose aggregates but also the enormous work involved in proving physical existence and membership by describing proper-motion and photometric work at Hamburg on the α Persei aggregate. For many years this cluster was known only as an aggregate of B and early A stars centered on α Per. No later type stars were believed to be present in the group. But a recent proper-motion study[3] in a field $5°$ square centered on α Per has now shown that the main sequence exists in this cluster extending at least as faint as G stars. It is impossible to pick out these faint members even statistically on direct plates because of the small ratio of cluster to field stars. Only by combination of proper-motion and spectral techniques can the true cluster population be found. Heckmann and Morgan emphasized the need for many studies of this kind to properly calibrate the absolute magnitudes of the B stars in loose aggregates. The need is to first get a finding list of possible loose clusters by the discovery technique employed by Stock and described by Nassau, Vyssotsky, and Lindblad, and then follow up with proper-motions, photometry, and spectral classification.

Although proper-motion data are essential to separate cluster members from field stars, it is often useful to make a statistical estimate of the number of superposed field stars to be expected in a color-magnitude diagram. This requires knowledge of the distribution function A (m, CI) where A is the number of stars per unit area at m in dm with colors CI in dCI. Parenago reported work by one of his collaborators, Mrs Starikova, on the computation of A (m, CI) from a generalized Schwarzschild equation

$$A (m, \text{CI}) = \omega \int_0^\infty D (r) \; \phi \; (M, \text{CI}) \; r^2 dr.$$

Parenago stated that comparison of the observed and computed A (m, CI) for any field provides a test for the presence of a true cluster.

(F) THE H-R DIAGRAM AND STELLAR EVOLUTION

(1) *The nearby stars*

Parenago discussed the H-R diagram for the nearby stars using spectral types and U-B-V colors for 851 stars with trigonometric parallaxes greater than 0″.03. Separation of the luminosity classes VI, V, IV and III was made on the diagram. Where estimates were not available, Parenago established the luminosity class from plots of spectral type against U-B. The separation was remarkably clear in these plots and showed that a powerful discriminate of absolute magnitude of a given spectral type is the U-B color. The classes VI (sub-dwarfs), V (main sequence), IV (sub-giants), and III (giants) appear to be separated on discrete sequences in the H-R diagram and Parenago interpreted these sequences as the result of stellar evolution by corpuscular radiation with consequent mass loss—a view which differs from the ideas of Schwarzschild and his school. Parenago pointed out that some of the sequences in the H-R diagram are running parallel to lines of constant radii and some are nearly perpendicular to these lines. He concluded that after an initial contraction from the inter-stellar medium a star can have two different types of development, one in which there is only a moderate change of radius as it evolves, and another in which a large change of radius is involved. No discussion was made of how these ideas apply to the systematic data available for star clusters, The final point which Parenago discussed was the identification of the main sequence in the globular clusters with the subdwarfs and high velocity giants of the nearby stars. With this identification, a fit of the C-M diagram of M 3 with the sub-dwarfs gives an absolute magnitude of + 1·5 for the RR Lyrae stars in M 3. The identification of certain field stars such as Arcturus and γ Leonis with the globular cluster sequence was also suggested.

(2) *Binary stars*

Binary systems provide unique information about certain aspects of cosmogony and stellar evolution. Members of a binary system are un-doubtedly of the same age but often differ in mass, absolute magnitude, and spectral type. This feature permits the study of evolutionary patterns in various stages of a star's complete history. Kulikovsky presented a summary of a study of 173 wide pairs. Spectral types, apparent magnitudes, and magnitude differences were used and the H-R diagrams for these pairs was obtained, but only with some difficulty because of the lack of data for many systems. Kulikovsky urged that interested observers should obtain photometric, spectroscopic and parallax data on many binary systems so that a complete study could be made. Many systems are easily accessible to observation but, strangely enough, are almost completely lacking in data. One typical example is π Andr with components of the 4th and 8th magnitude separated by 36 sec of arc. The spectral type and colors are unknown for the companion. According to Kulikovsky twenty such easy cases exist within reach of only moderate instruments. Following a recommendation of Commission 26 of the I.A.U., Kulikovsky suggested that observers with large instruments should undertake spectral classification, photometry, and parallax determinations for a list of binary stars selected to be of the greatest interest. Kulikovsky is willing to prepare such a list and will communicate this to any interested observer. Parallax observations are especially desired and the parallax observers at the Yerkes, Dearborn, Swarthmore, and Leander-McCormick observatories should be alerted.

Bidelman at Lick has a large spectral classification program under way for the components of binary stars and Oke and Bakos at the David Dunlap Observatory have undertaken photo-electric photometry on wide pairs to obtain a C-M diagram. Both these efforts are along the lines suggested by Kulikovsky. Co-ordination of effort may be desirable. If so, it can best be accomplished by private correspondence between the interested parties.

(G) ASSOCIATIONS

Kharadze stated that no observatory has a regular program for the study of spectral characteristics of stars located in regions of large diffuse hydrogen nebulae. These stars have been discovered and investigated by A. H. Shajn in associations and star chains. At Abastumani Observatory spectral classification and spectro-photometry has begun for these stars with a new meniscus telescope of 70 cm aperture with an objective prism of the same size.

Flare and flash stars are common in T associations. But the question of their general space distribution is not settled. There are at least two schools of thought. Kholopov believes that flare and flash stars are distributed uniformly over the sky, while Haro's experience suggests that they are concentrated in T associations. Haro also suspects that there is a correlation of the characteristic flare time with the type of association in which the stars are located. For example, in the Orion association the flare time from minimum through maximum and back to minimum is about 120 minutes according to the Tonantzintla observations. In the S Mon association (NGC 2264) the characteristic time is about 60 minutes according to Haro's data and about 40 minutes according to Rosino. In the Taurus dark clouds, the time is only 28 minutes. This is obviously a new and important field in which much work remains to be done. The problem is important from the standpoint of the initial origin of stars because M. F. Walker's observations of C-M diagrams of young associations show that T Tauri stars (and some flare stars?) are probably in the stage of gravitational contraction to the main sequence and may represent the earliest stage in the life of a star which can be directly observed. Haro emphasized the need for detailed work on this problem and stated that co-ordination could be made through the committee on problems of T Tauri stars composed of Herbig, Haro, and Kholopov, which was appointed at the Burakan conference in 1956. All interested observers with suitable equipment should contact this committee concerning co-ordination problems.

References

[1] See, for instance, Sandage, A. *Atroph. J.* **125**, 435, 1957.
[2] Trumpler, R. J. *Lick Obs. Bull.* **18**, 167, 1938.
[3] Heckmann, O. and Lübeck, K. *Mitteil. Astr. Ges. 1956*, **16**, 1957; see also *Z. f. Aph.* **45**, 243, 1958.

II. THE GALAXY

(A) INTRODUCTION ON THE IMPORTANCE OF STELLAR
EVOLUTION FOR PROBLEMS OF GALACTIC STRUCTURE

In connexion with the importance of present views on stellar evolution for discussions of the evolution of the Galaxy, the sessions dealing with galactic problems were preceded by a review of items relating the evolution of stars with that of the Galactic System. The following represents with only few modifications the introductory paper on this subject as presented by J. H. Oort.

Although we are still far from understanding how the galactic system has evolved, and how stars composing it have been formed, it seems that we have nevertheless approached a stage where it becomes useful to go beyond a simple description of stellar distribution and motions. Gradually some facts are emerging, which suggest at least a rudimentary understanding of why different objects are distributed in the way they are distributed.

A theory of the galactic system is primarily a theory of its evolution. Before considering this we summarize the principal facts concerning the differences between various constituents of the Galaxy. An extensive list of such differences has been given by Parenago at the Groningen conference in 1953[1]. A more limited compilation is presented by Table 2. The sub-divisions and the names of the various groups are those advocated by the participants of the *Semaine d'Etude* on Stellar Populations (Rome, May 1957). For each population group rough indications are given of the average distance, z, in parsecs from the galactic plane; of the average speed, Z, in km/sec, in the direction perpendicular to the galactic plane; of the axis ratio of the sub-system; of the degree of concentration toward the galactic centre; and of the smoothness of the distribution. Next we give estimates of the heavy element abundance for each population group, of its age and of its total mass.

At the extreme left of Table 2 are the stars that populate the spherical sub-system, called the halo population II. Examples of objects belonging to this population are the sub-dwarfs, which probably form the most massive part of the population, the typical globular clusters, and the RR Lyrae variables with periods longer than 0·4 days approximately.

The second column represents an intermediate population II. The long-period variables of early spectral classes and relatively short periods belong

Table 2. *Subdivision of objects in the Galaxy into population groups*

Population	Halo population II	Intermediate population II	Disk population		Older population I	Extreme population I
			Planet. nebulae, bright red giants, novae	Weak-line stars		
Typical members	Sub-dwarfs, glob. clusters, RR Lyrae var. with period $> 0\overset{d}{\cdot}4$	High-vel. F-M stars, long-period variables	Planet. nebulae, bright red giants, novae	Weak-line stars	Strong-line stars, A stars, Me dwarfs	Gas, super-giants, T Tauri stars
z (pc)	2000	700	450	300	160	120
Z (km/sec)	75	25	18	15	10	8
Axial ratio of sub-system	2	5	$\simeq 25$	—	—	100
Concentration toward centre	Strong	Strong	Strong	?	Little	Little
Distribution	Smooth	Smooth	Smooth	?	Patchy, spiral arms	Extremely patchy, spiral arms
Heavy element content (Schwarzschild)	0·003	0·01	—	0·02	0·03	0·04
Age (10^9 years)	6	6·0–5·0	5	1·5–5	0·1–1·5	<0·1
Total mass ($10^9 \odot$)	16	47			5	2

at least partly to this intermediate population; also the ordinary high-velocity stars.

We next have the disk population, which has been divided over two columns in Table 2. Examples of this population are the planetary nebulae and novae, which have a rather strong concentration to the galactic plane. As another part of this disk population we consider the category of weak-line stars of Miss Roman. While it is well known that the objects of the first three columns show strong concentration towards the galactic centre, nothing is known about the distribution of the weak-line stars in the plane of the Galaxy.

The intermediate population I or "Older Population I", like class A stars and the strong-line stars can be clearly distinguished from the rest. The same holds for the extreme population I, which includes among other objects the super-giants and the interstellar gas.

The estimates of the total mass of the various population groups, given in the bottom line of Table 2, are extremely uncertain and have been included only to give an indication of the relative mass distribution among the different components of the Galaxy. The most certain of these estimates is that of the extreme population I, based on the 21 cm measures of the interstellar hydrogen. For the intermediate population I the estimate has been taken from Schmidt's[2] model computation; this value must be considered quite uncertain. The mass of the halo population II has been obtained from the number of sub-dwarfs in the vicinity of the sun, assuming that the sub-dwarfs and the globular clusters of the halo population have the same distribution over the Galaxy. This estimate is also very uncertain, because we certainly do not know all the sub-dwarfs in our vicinity and therefore this figure represents a lower limit. If the mass values given for the three population groups mentioned are subtracted from Schmidt's estimate of the total mass of the galactic system, there remains 47×10^9 solar masses. Most of this is probably the disk population, with a small contribution from the intermediate population II.

We may imagine the Galactic System to have resulted from the contraction of an irregular mass of gas. This mass of gas may have been endowed with a considerable resultant angular momentum, which will still be largely present in the galactic system. Some of it may have been carried away by escaping stars or masses of gas, but it seems likely that most of it is still present. From estimates of the total angular momentum of the Galactic System we may therefore arrive at an estimate of the total angular momentum that was present in the primeval mass of gas.

The way in which this mass of gas developed would depend in the first place on the proportion of regular and irregular motion. If the motions were quite regular this primeval gas would have contracted in a thin disk of approximately constant density. On the other hand, if there was only a small amount of angular momentum present, the primeval gas would have contracted into an almost spherical system. The amount of concentration to its centre would depend on two factors: on the rate of star formation during the various stages of the contraction and on the random motions of the clouds from which the stars were formed. It is evident that the galactic system is something between these two extremes. Its mass has certainly a flattened shape, but at the same time the main mass is strongly concentrated to the centre.

The halo population II, in the first column of Table 2, should then contain the oldest objects, formed before the contraction into the disk. The measure of central concentration of this halo population should thus give important information on the original internal motions in the gas.

It is probable that during the process of contraction to a disk the internal currents of the gas formed layers of high density from which a large number of stars were formed. It is difficult to make an estimate of the time which this contraction will have taken. The age of the halo population II seems to be of the order of 6×10^9 years as derived from the colour-magnitude diagrams of globular clusters. The time of contraction to the disk, during which the intermediate population II was formed, will depend on the character of the regular and irregular streamings that were present in the gas, and on the amount of energy radiated away by the heated hydrogen gas. This heat had to be radiated away in order to make further contraction possible and this puts a limit on the time of contraction. A very rough guess is that it would have taken about 10^9 years for the system to contract from the original stage to the disk stage. After the contraction to the disk had been completed the disk population must have been formed. This is probably the major part of the population of the Galactic System.

Since there must have been a very strong concentration of matter towards the centre by this process, it is not surprising that the disk population exhibits such a concentration, but we may ask why the gas at present shows no concentration at all towards the centre. According to the observations it is rather evenly distributed over the disk, perhaps even somewhat less dense in the central part than in our neighbourhood.

To explain this difference we are probably forced to assume that the star-forming process depends on a fairly high power of the density. If this is so, we may imagine that stars were formed during all stages of the contraction, but that the star-building process was particularly rapid in the last stages when matter had become very dense in the inner parts of the disk. In the following, and presumably longest, stage in the existence of the Galactic System star-forming would have gone on more slowly and at a gradually decreasing rate.

If there is any truth in the above sketch of the evolution of the Galactic System the classification given in the table must be roughly a classification by age.

The distinctions in the last columns of Table 2 are of a gradual kind. The very youngest objects and the gas are in the last column, and objects of which we have some indication that they are fairly young have been put in the column next to it. It is difficult to assign ages to objects in the disk population, except in the case of the galactic clusters. The difficulty in disentangling ages of objects in the post-contraction period is due largely to the fact that apparently the dynamical properties of stars formed at different times between 5 and 0·5 billion years ago have changed very little. There are, however, indications that the O and B stars and the galactic clusters containing stars of early spectral class have average distances from the galactic plane of only 50 or 60 pc, while we find twice these distances for the clusters with ages about 5×10^9 years. This difference may be supposed to be due to the fact that gradually the turbulence of the interstellar gas has been decreasing. It seems, however, more probable that stars and clusters are still being accelerated by encounters with large irregular concentrations of interstellar matter. An ideal way of investigating this last problem would be, as was once tried by Weaver[3], to study galactic clusters, which according to the H-R diagram have different ages, and to see if one can find an increase with age of the average distance from the galactic plane or of the average velocity perpendicular to the plane. There seems to be a good indication that such is the case. One could even from these properties estimate the total effect that encounters with large interstellar masses have had on the velocity distribution.

Some important aspects of the structure of the galactic system are still entirely missing in the sketch given above. It does not explain how the spiral structure, which seems to be such a very common feature of galaxies, has come into existence. The same holds for the bar structure which seems to be so common an element in galaxies.

In recent years more and more evidence has accumulated that stars formed at different epochs have different chemical composition. The fraction of elements heavier than H and He appears to increase from about 0·003 in the oldest stars to about 0·04 in the youngest ones. The change is probably connected with a gradual change with time in the composition of the interstellar medium from which the stars were formed. If this is so, the enrichment with heavier elements must have been due to transformation of elements inside stars which have subsequently shed a large part of their mass into the interstellar medium again. It is then quite possible that all heavier elements now present in the interstellar medium have been formed in stars. It is still quite uncertain whether the enrichment is due to the explosion of a relatively small number of super-novae or whether the larger part comes from slower processes of expulsion of matter from stars.

Table 2 contains estimates by Schwarzschild [4] of the proportion of elements other than hydrogen and helium in the stars of various population groups. From the fact that some globular clusters at large distances from the galactic plane give evidence of a considerable metal content, while others show hardly any metal lines, and that, as Morgan has pointed out, a number of globular clusters near the galactic nucleus show heavy-element abundances resembling those in the sun, it would appear that the major part of the enrichment with heavier elements has occurred in an early phase of galactic evolution, part of it even before the Galactic System had contracted into a disk.

A development of special interest in the study of galactic structure has recently been given by Salpeter. He remarked that if we suppose that in our vicinity the star-forming process has proceeded at roughly the same rate during the past 4 or 5×10^9 years a plausible form of the generating function (i.e. the luminosity law of stars formed at any given time) would explain the present somewhat peculiar shape of the luminosity function, in particular its rather sudden bend near $M = +3$. The luminosity laws observed in galactic clusters, which resemble Salpeter's generating function, lend support to this idea, as Sandage has shown. This indicates, moreover, that the generating function may have been approximately the same in clusters and in the Galactic System at large. Salpeter has indicated that if this plausible idea of a continuous birth of stars is accepted and if we assume, further, that all the stars brighter than about absolute magnitude $+3$, which have burnt up the hydrogen in their cores, have become white dwarfs, the number of white dwarfs in our neighbourhood must be about 10 % of the total number of stars on the main sequence, while in our region of the Galactic System the mass of gas expelled by the stars that are now white dwarfs must have been of the same order as that of all stars or about three or four times that of the interstellar gas now present. This implies that the gas must probably have passed several times through stars.

The appearance of a star is determined largely, if not entirely, by *two* parameters, its *mass* and its *content of heavy elements*. If the enrichment with heavy elements has occurred at the same rate in all parts of the Galactic System the heavy-element content of a star would also determine the epoch at which it was formed and, therefore, its age. If this were so and if the heavy-element content of a star could be determined with sufficient accuracy, we might in this way arrange the stars in age groups. In practice we cannot yet reach this ideal. In the first place we can as yet hardly measure the heavy-element content of an individual star except when it deviates drastically from normal, as in the case of the sub-dwarfs. In the second place it is likely that the enrichment of the interstellar material has depended on the amount of star formation and has therefore been much greater in the central parts than in the outer regions.

Attempts to determine the heavy-element content with greater precision must be considered to be of the greatest importance. It is probable, as Sandage has shown, that considerable progress in this line can be made by studying galactic clusters. The main difficulty *there* is that we know as yet so few clusters of ages intermediate between the primeval and contracting stage on one hand, and the youngest stage of the System on the other hand.

References

[1] For particulars, see *Astronomical Newsletter*, no. 71, 10, 1953 and no. 73, 2, 1954.
[2] Schmidt, M. *Bull. Astr. Inst. Netherl.* **13**, 15, 1956.
[3] Weaver, H. F., unpublished.
[4] See the forthcoming report on the *Semaine d'Etude* on Stellar Populations, Vatican, 1957, now being edited by D. O'Connell.

(B) SOME PROPERTIES OF OTHER STELLAR SYSTEMS

This section deals with a few remarks on stellar systems outside the Galaxy which were considered to be of importance in the interpretation of galactic observations.

(1) *Integrated spectra and colours of galaxies*

Information on the general stellar population predominant in the light emitted by galaxies, may be obtained from a study of the integrated spectra. W. W. Morgan reported on such an approach, recently made by Morgan and Mayall [1], extending earlier work by Humason [2] and others.

The criteria used for the spectral classification, which necessarily are of the very low-dispersion variety, are mainly the appearance of the hydrogen lines and the cyanogen absorptions at $\lambda\lambda$ 3800 and 4200. In objects where the hydrogen lines are of maximum intensity the predominant stellar population is composed of stars of about spectral class Ao, while in objects showing maximum strength of the CN absorption the predominant stars are of spectral classes somewhat later than Ko. Composite spectra between these two extremes may be evaluated in about the same way as an unresolved binary is classified.

A number of the brighter galaxies have been classified and a marked correlation between the spectral class and the nebular type according to the Hubble classification appears. This was also indicated in the work of Humason. When making comparisons between various galaxies it is important, however, that only spectrograms covering the same wave-length region are used.

The red and infra-red regions of the spectrum of M 31 show strong TiO bands, indicating that a large proportion of the luminosity is due to M stars, which are probably similar to the many M stars found in our Galaxy by Nassau and his collaborators. In the blue region the CN absorption is prominent, indicating that also a large number of giant K stars is present. Other galaxies of the Sb type with a large central bulge have spectra of the same type as M 31. The irregular galaxies, such as NGC 4449, and the Sc systems, which have no marked central condensation, have spectra of early type; most of the luminosity is in this case due to stars around spectral class Ao. The giant ellipticals, like M 87, have an almost pure K spectrum, while dwarf ellipticals, such as M 32, are of earlier spectral class and somewhat peculiar.

Observations of M 31 reveal no change of the population with distance from the nucleus. On the other hand, the colour indices as determined by Holmberg[3] show considerable variation from the inner to the outer regions. In the central part the colour index is always large, about $+0.84$, the same as for the elliptical galaxies, while in the outer parts of the spiral it amounts to between $+0.40$ and $+0.50$. The irregular galaxies according to Holmberg seem to belong to two different types, one having a mean colour index of $+0.28$, whereas the other is the same as for the ellipticals.

An especially interesting case is M 82, which exhibits a large colour index but a pure class A spectrum. It belongs to the turbulent type of galaxies which in a number of cases are radio sources. They have strong emission lines, as for instance NGC 4151. Perhaps the emission lines originate in a very large number of planetary nebulae, not in H II regions.

(2) *The rotation curve of M 31*

In order to compare the 21 cm rotation curve of the Andromeda nebula with that of our Galaxy van de Hulst and Raimond [4] have determined 21-cm profiles at various points on the major axis of M 31 out to 2° from the centre on either side. Westerhout reported on this work. Van de Hulst has computed the density distribution and the rotation curve fitting the observations. The resulting rotation curve, which shows very little asymmetry, does not drop as fast in the outer regions as does the rotation curve of our Galaxy based on Schmidt's model. This may indicate that the mass density in the outer regions of the Galaxy has been underestimated.

The mass density of the neutral hydrogen in M 31 shows maxima at about 1° from the centre, that is where the main spiral arms appear and where the super-giants are most conspicuous. Similar maxima are shown in the Galaxy at distances of about 7 kpc from the centre. There is a difference between the two systems in that M 31 shows a long tail of low density stretching out farther in the equatorial plane than is the case in our Galaxy, where the density drops fairly rapidly.

As mentioned above the radio measures of M 31 show very little asymmetry of the rotation curve, of the order of 15 km/sec. In contrast to this the emission nebulae have given a difference amounting to about 100 km/sec between the south preceding and the north following side. It was suggested by Lindblad that this might be due to M 31 being a barred spiral, but it is difficult to see why the neutral and the ionized hydrogen should then behave differently. Baade thought that the observed difference is due to tidal disturbances from M 32, and pointed out that part of it might be introduced if the emission nebulae are not all situated in the equatorial plane of the system. Fehrenbach stressed the difficulty of measuring exact radial velocities of emission nebulae and recommended the use of interference methods. He intends to use this method at Haute Provence to check the previous measurements.

References

[1] Morgan, W. W. and Mayall, N. U. *Pub. Astr. Soc. Pacif.* **69**, 291, 1957.
[2] Humason, M. L. *Astroph. J.* **83**, 18, 1936.
[3] Holmberg, E. *Medd. Lund Astr. Obs.* Ser. II, no. 136, 1958.
[4] Hulst, H. C. van de, Raimond, E. and Woerden, H. van. *Bull. Astr. Inst. Netherl.* **14.** 1, 1957.

(C) THE GALACTIC HALO

The objects from which the structure of the galactic halo may be derived by optical means are the extreme population II objects such as RR Lyrae variables with periods longer than 0·4 days, the sub-dwarfs, globular clusters, stars similar to those found in globular clusters such as the bright red giants, and the blue stars found in the galactic polar regions. The long-period variables with periods shorter than 200 days and the ordinary high-velocity stars are now regarded as belonging to an intermediate population II, somewhat younger than the halo population II proper.

New important information on the halo of our Galaxy and of some other galaxies has in recent years been obtained from the radio-astronomical observations. As the optical and the radio approaches are quite different they are discussed separately below.

(1) *Optical work*

The Palomar-Groningen variable star survey

From the Groningen conference emerged the programme for finding the general distribution of the variable stars in the galactic halo[1]. Three fields have been chosen at the longitude of the galactic centre and at various latitudes and one in Cygnus. An important criterion for the selection of these regions was uniformity of the absorption in the fields if any absorption at all. From the first three fields we should get information on the density distribution in a meridian plane of the galactic system through the sun and the centre, and from the fourth field data on the distribution in a plane perpendicular to this meridian plane at about the same distance from the galactic centre as that of the sun.

l	b
$327°5$	$+28°$
$331·0$	$+12$
$327·5$	-12
$48·0$	$+10$

Dr Plaut visited the Mount Wilson and Palomar observatories in 1956 and obtained of each field 100 photographic plates and twenty photovisual ones, using the 48-inch Palomar Schmidt. The size of the regions to be investigated is $6°6 \times 6°6$. The exposures are such that the plate limit is below the 20th photographic magnitude, and the survey aims to completeness down to 17·5 median magnitude, corrected for absorption.

Plates have also been taken by Dr Plaut for the transfer of the photometric system from the Selected Areas observed by Baum to the regions in question. It is not yet known if this photographic transfer will meet all the requirements. Eventually, photo-electric sequences in the fields themselves will be desirable.

It was estimated already during the Groningen conference that the work involved in the finding of the variables and in the reductions would be considerable; the estimate was about one man full time for about six years. Since then an effort has been made to reduce the time required for finding the variables, and an electronic device has been developed at the Kapteyn Laboratory by Borgman in which the blinking is made semi-automatically. A television screen shows the difference in the brightness of the stars of the two plates compared. A difference in brightness of a star shows as a light or dark spot on the screen. Detection of the variables thus becomes easier and more variables with small amplitudes will be found than in the traditional blinking procedure. Preliminary tests indicate that the semi-automatic method is indeed more efficient than the visual method. The regular scanning for the variables will probably have started at the end of 1957.

It thus seems that this programme, which was initiated four years ago, will proceed as it was hoped, and may produce valuable information on the short-period variables in the first place, and later also on the distribution of the long-period variables. For this latter purpose plates are still being taken. Addition of a fifth field in the galactic polar region is contemplated.

It was emphasized during the conference that it would be extremely valuable to get accurate photo-electric colours for the RR Lyrae variables which will be discovered in this survey. As the stars in question will be very faint the problem will be a difficult one, which only can be tackled by few observatories.

Other variable star surveys

Other surveys of variable stars which may add information on the galactic halo are in progress at several observatories, for instance at Leiden, Stalinabad, Sonneberg, Bamberg, Odessa and Moscow. The magnitude limits of these surveys are generally brighter than in the case of the Palomar-Groningen survey, but they cover larger areas. All unstudied variables brighter than 12th magnitude at maximum are investigated at Stalinabad. New variable stars brighter than 12 to 13th magnitude are being searched at Bamberg. The Soviet observatories are proceeding with a study of variables down to the 16th and 17th magnitude, particularly in the northern Selected Areas. An extension of this programme is planned within four or five years, as soon as the new 104-inch reflector of the Crimean Observatory and the 40-inch Schmidt telescope of Burakan Observatory come into operation.

Dr Kurkarkin in this connexion repeated the offer, already made at the Groningen conference, that the Soviet astronomers are ready to co-operate in measuring and investigating plates from observatories which do not find time to evaluate the plates themselves.

Search for faint blue stars

Following another recommendation made at the Groningen conference, the Tonantzintla Observatory is engaged in a search in the galactic polar caps for faint blue stars of the type discovered by Zwicky and Humason. With the Schmidt telescope three exposures are made on each plate (Kodak 103 a-D) using successively ultra-violet, blue, and yellow filters. There are thus three images of each star, exposed in such a way that their densities are the same for a star of spectral class A5. By this technique it is easy to pick out the blue stars.

The limiting photographic magnitude of the present survey is about 17. Approximately 1360 square degrees had been covered in the northern polar cap up to the time of the conference, in which about 850 new blue stars were found. If we add to these the similar stars discovered by Zwicky, Humason and Luyten, which number about 400, we have nearly one such blue star per square degree. A similar survey is in progress in the southern polar cap, and in co-operation with Dr Luyten similar work is planned to the 19th photographic magnitude in some specific Selected Areas by means of the Palomar Schmidt.

The finding list of the blue stars in the northern polar cap has been published [2] and the list of objects in the southern cap will probably be ready by the end of 1957. By means of these finding lists the objects may be identified by other observers for photo-electric photometry and for slit spectroscopy.

The blue stars discovered at Tonantzintla have been plotted together with the similar stars previously known. The resulting apparent distribution on the sky shows some irregularities, but most of these may be due to the inclusion of the stars found by Zwicky, Humason, and Luyten. Some of these do not fulfil the Tonantzintla criterion for selection; when they are disregarded a relatively smooth distribution is obtained. No detailed statistics have been made as yet, but it is obvious that the number of blue stars in the polar caps increases rapidly with decreasing apparent brightness.

Some of the extremely blue stars are perhaps variables of the SS Cygni type. In order to establish this more plates will be taken.

(2) *Radio observations*

After his discovery of the halo of M 31, Baldwin[3] four years ago made a galactic survey at 3·7 m. He found that the radiation is not homogeneously distributed over the sky but markedly concentrated around the direction of the galactic centre. This was also found in the surveys of Bolton and Westfold[4] at 3 m, and in those of Dröge and Priester[5] at 1·5 m. These observations prove the galactic origin of most of the radiation.

Recent surveys by Mills[6] at 3·5 m also show the halo component very well, but it is somewhat difficult to separate it from the extra-galactic component. According to Mills the halo component has an angular width between half-brightness levels of 60° to 70°, which corresponds to 10 kpc in the region of the galactic centre. Baldwin has computed a number of models fitting his observations. They have the form of an almost spherical sub-system (axial ratio > 0·7) with uniform emission per unit volume and a radius of about 14 kpc.

A survey of 21-cm profiles at high galactic latitudes was made at Kootwijk[7] in 1955. The profiles show high tops going up to 20° K at about zero radial velocity and very long wings of low intensity. The average extreme velocities of the wings are − 42 and + 24 km/sec, and there does not seem to be any correlation between the observed velocity values and the direction. Slightly more negative than positive velocities seem to be present. In some profiles recently obtained at Dwingeloo, the central maximum is much wider in velocity than the maxima observed near the galactic equator, with half-widths corrected for the band-width of the order of 30–50 km sec. The neutral hydrogen in the halo is thus in a much more turbulent state of motion than that in the galactic plane.

Although the density must be very low the continuum observations seem to indicate that the halo, owing to the large volume, contributes a fairly substantial proportion of the whole galactic mass. At 3·5 m the halo accounts for something like 90% of the continuous radiation from our Galaxy. The main difficulty for a more thorough evaluation is the superposed extra-galactic component. In principle the two components may be separated, because the halo component is spheroidal around the centre while the extra-galactic component should be spherical around the sun, but in practice the separation is very difficult.

Another possible way of evaluating the general extra-galactic contribution has recently been suggested by Shain[8]. In low frequencies the ionized hydrogen regions will be opaque and blot out all extra-galactic radiation. Hence, if we observe a distant H II region in high galactic latitude, for instance the nebula 30 Doradus at, say, 15 m, we will measure only the radiation from the gas in front of the nebula. From comparisons with nearby regions we may then evaluate the radiation from the background.

At radio-frequencies the halo of our Galaxy shows less concentration to the centre than the globular clusters and other typical halo population II objects. As early type galaxies, rich in population II, are not detected as radio emitters it seems that the radio halo is not immediately connected with population II [9]. The halo observed at radio frequencies seems to consist partly of neutral hydrogen, and partly of high energy electrons radiating according to the synchrotron mechanism as suggested by Sklovsky [10]. The general distribution, the non-thermal spectrum, and the lack of fine-structure indicate this latter mechanism.

The most complete observations of a halo outside the Galaxy are those of M 31. Baldwin [11], at 3·7 m, finds a rather smooth distribution, no central maximum and a diameter between half-intensity points of 2°4. At 75 cm, Seeger, Westerhout, and Conway [12] find radiation out to 5° from the centre, having a very flat distribution with half-width 6° × 3°5. After correcting for the known tilt of M 31 one finds a value of 3 : 1 for the ellipticity of the halo, which is quite large compared with Baldwin's estimate [3] for the galactic system. At 21 cm, van de Hulst, Raimond, and van Woerden [13] have detected neutral hydrogen out to 2°5 from the centre on the major axis of M 31.

According to observations at Sydney, the Magellanic Clouds at 3·5 m show very little sign of a halo [6]; the whole radiation is concentrated to the central part of these systems. At 21 cm, however, an extensive halo of neutral hydrogen is observed [14]. This is contrary to the halo of our Galaxy, which is most important in the continuum.

NGC 5128 has very recently been shown by Mills and Sheridan at 3·5 m to be a very large elongated object with a major axis of 8° to 9°, much larger than the optical object, which is about 25′ in size. Here also we have a gigantic halo, but it is very flattened.

Finally, recent 21-cm observations by Heeschen of M 51 and M 81 also show the existence of large haloes of neutral hydrogen. The radio diameter of M 51 is about 2° as compared with the optical diameter of about 15′. The observations of M 81 are less certain, but a large halo is certainly indicated.

During the conference Oort expressed some doubt as to whether the

21-cm observations of other galaxies far from the central regions refer to matter in the haloes. It may well be disk matter. From the evidence of the neutral hydrogen in our Galaxy at high galactic latitude one would expect a large velocity dispersion in the haloes of other galaxies as well. This is not shown by the present observations.

References

[1] Symposium No. 1 of the I.A.U. *Co-ordination of Galactic Research*, ed. A. Blaauw (Cambridge University Press, 1955), p. 4.
[2] Iriarte, B. and Chavira, E. *Bol. Obs. Tonantzintla y Tacubaya*, no. 16, 3, 1957.
[3] Baldwin, J. E. Symposium No. 4 of the I.A.U. *Radio Astronomy*, ed. H. C. van de Hulst (Cambridge University Press, 1957), p. 233.
[4] Bolton, J. E. and Westfold, K. C. *Austr. J. Sci. Res.* A3, 19, 1950.
[5] Dröge, F. and Priester, W. *Z. f. Astroph.* 40, 236, 1956.
[6] Mills, B. Y. *Austr. J. Phys.* 8, 368, 1955.
[7] Westerhout, G. Symposium No. 4 of the I.A.U. *Radio Astronomy*, ed. H. C. van de Hulst (Cambridge University Press, 1957), p. 26.
[8] Shain, C. A. *Mt Stromlo Pre-conference on Co-ordination of Galactic Research*, p. 53, Canberra, 1957.
[9] Mills, B. Y. *Radio Frequency Radiation from External Galaxies*, Handbuch der Physik, in press.
[10] Sklovsky, I. S. *Dok. Akad. Nauk S.S.S.R.* 90, 983, 1953.
[11] Baldwin, J. E. *Nature*, 174, 320, 1954.
[12] Seeger, Ch. L., Westerhout, G. and Conway, R. G. *Astroph. J.* 126, 585, 1957.
[13] Hulst, H. C. van de, Raimond, E. and Woerden, H. van. *Bull. Astr. Inst. Netherl.* 14, 1, 1957.
[14] Kerr, F. J., Hindmann, J. V. and Robinson, B. J. *Austr. J. Phys.* 7, 297, 1954.

(D) THE NUCLEAR REGION

The nuclear region may be explored by means of both radio and optical data. Optical methods include surveys of variable stars and various other objects, but of special interest are measurements in the far infra-red, which may provide a connexion between optical and radio observations.

(1) *Optical investigations*

Variable stars

Surveys of variable stars in three selected fields in the nuclear region were reported by Oosterhoff. The centres of the fields are

	l	b
Sgr I	$329°.1$	$-4°.0$
Sgr II	331.7	-6.6
CPD $-31° 5547$	329	-10

The first two fields were selected by Baade to supplement the investigation of a previously selected field in the central region, the results of which have recently been published by Gaposchkin [1]. Most of the plates of these two fields, which cover areas of 0°60 × 0°45 with limiting magnitude 19, were taken with the Radcliffe reflector. Some plates of Sgr II were taken by Baade with the 100-inch reflector at Mount Wilson. The third field, investigated by means of the Franklin-Adams camera, covers 10° × 10° and has limiting magnitude about 16·0. The total number of plates of the three fields are about 90, 110, and 300, respectively. The investigation of the plates was started by the late Mr Schuurman and has been continued by Mr Ponsen.

Although only part of the plates have been blinked as yet 115 variables have been found in the field Sgr I. This number is probably approximately complete, and for the eighty-one variables which have been estimated the distribution over different types is as follows:

RR Lyrae type	30
Long-period	22
Semi-regular	17
Eclipsing binaries	4
Uncertain	8

The last group, which will be further investigated, probably contains a few Cepheids. During the further investigation the relative number of RR Lyrae stars and semi-regular variables may well increase considerably, as many of them have small amplitudes. Until now only few periods have been derived; they confirm the well known tendency of the long-period variables in the nuclear region to show periods below 250 days, the majority may perhaps even have periods below 150 days. As for the RR Lyrae variables, both periods of 0·3 day and of 0·5 day and longer are represented. No light-curves are available as yet. From a first inspection of the plates, the interstellar absorption in Sgr I seems far from homogeneous.

In the field Sgr II, situated somewhat further from the galactic centre, about fifty variables have been found, but there are probably some more to be discovered. Forty-eight variables have been classified; they are distributed as follows:

RR Lyrae type	23
Long-period and semi-regular	20
Eclipsing binaries	1
Uncertain	4

For five long-period and semi-regular variables and for twenty-one RR Lyrae variables periods could be derived. Only one star in the long-period group has a period longer than 250 days; in the RR Lyrae group the majority has periods about 0·5 day or longer, but there are several with periods around 0·3 day.

Both in Sgr I and II the frequency maximum of the RR Lyrae stars is found at about magnitude 16 or 17. The same thing was very pronounced in Baade's field close to the centre[1]. In all three cases it is therefore to be assumed that the variables really belong to the nuclear region and that the investigations have penetrated the central bulge. The nature of the variables seems to change between the three fields. In Baade's field a large number of very short-period RR Lyrae variables were found (periods below 0·4 day). Even if some of the periods are spurious this fact seems to be established. In the field Sgr II the periods of the RR Lyrae variables are longer, and probably Sgr I will prove an intermediate case in this respect.

In the third field mentioned above, CPD -31°9547, which has much larger field, a systematic survey for long-period variables is in progress. The first plate pair yielded 146 variables of this type, the majority of them being new objects. 90 % of these stars has maximum brightness between 13 and 14·5 photographic magnitude. The number of RR Lyrae variables is only nine, owing to their lower absolute magnitude. There is little doubt that the numerous long-period and semi-regular stars belong to the outskirts of the central galactic bulge. Although it is difficult to give an accurate estimate, the total number or these stars in the area is considered by Ponsen to be about 1000, thus forming valuable material for a statistical investigation.

In this connexion it is of interest that Oosterhoff[2] has recently investigated the population characteristics of the Cepheids in a region 70° × 70° around the nucleus. Most of the Cepheids in this area are situated south of the galactic equator, a distribution which is quite different from that of other stars. In order to make a statistical separation of Cepheids of population I and population II the frequency distribution of the periods was studied, and from comparisons with statistics for pure samples of the two populations it was found that the Cepheids brighter than magnitude 10 or 12 within the belt ±4° latitude give a frequency distribution quite similar to that of the normal population I Cepheids. For all the others the frequency distribution is nearly the same as that of the Cepheids in the globular clusters. The number of Cepheids studied was about 120 and it thus seems that about forty of the brighter objects are population I Cepheids probably associated with a spiral arm in the direction of the centre, while the others are typical population II objects scattered around the galactic nucleus.

The infra-red spectral surveys reported by Nassau have disclosed a large number of red variables in the nuclear region. The area around the globular cluster NGC 6522 has been especially studied, partly by means of photometric material provided by Baade and by Haro. The frequency maximum of the variables is found at about 12th magnitude infra-red. Most of the stars appear to be of rather late spectral class and the majority probably belong to the semi-regular or irregular type of variables.

If there is a clustering of variable stars of various types in the galactic nucleus, as seems probable, it is of greatest importance to investigate the magnitude differences, thus the differences in absolute magnitudes, between the irregular and semi-regular variables on the one hand and the long-period variables with relatively short periods and the RR Lyrae stars on the other.

Survey of M giants

The Cleveland infra-red spectral surveys show that the non-variable M stars are distributed in about the same way as the red variables. In the nuclear region there appear to be relatively more stars of late spectral classes than in other regions investigated.

Planetary nebulae and novae

As is well known the planetary nebulae and novae show a concentration towards the galactic nucleus. However, according to Baade the concentration towards the disk of the Galaxy is more significant and the distribution of these objects is therefore discussed in the section dealing with the disk of the Galaxy.

Infra-red continuum surveys

Surveys of the infra-red radiation at a wave-length of about 1μ from the nuclear region have been made by Stebbins and Whitford[3], by Kaliniak, Krassovsky, and Nikonov[4] and by Dufay and his collaborators[5]. The results seem, however, still to be mixed up with absorption effects to an extent that it is difficult to distinguish between absorption and real structure. In order to obtain more definitive results it is necessary to go further into the infra-red than has been done so far. It was reported at the conference that experiments at about 15μ are in progress at the Naval Research Laboratory in Washington. Whitford was also reported to have started work at about 10μ. There is an entirely clear window in the atmosphere in this region, but it is very difficult to construct receivers which are sensitive enough. These infra-red surveys are important also because they provide a bridge between the optical and radio investigations.

(2) *Radio Observations*

Until recently most of the continuum surveys of the nuclear region have been made with aerials having beam-widths larger than 8°. Westerhout reported that only six surveys have been published which give enough detail to distinguish discrete sources in this region. In addition, two single observations have been made at very short wave-lengths. At these short wave-lengths, the main feature in the direction of the galactic centre is a discrete source with a diameter of $0°25$ to $0°50$, designated as source no. 5 by Haddock, Mayer and Sloanaker[6], which is superposed on a broader source with a diameter of about $2°$, designated no. 9. The combination of these two sources is clearly different from the underlying galactic ridge, which has roughly the same width and intensity over a region covering a longitude interval of $40°$ around the centre.

At the longer wave-lengths the situation is more confused, as the galactic ridge begins to bulge out in latitude, and also the intensity of the ridge rises slowly towards the centre. However, from Mill's survey[7] at 3·5 m it is evident that in the direction of the centre a source having the width of about $3°$ is superposed on the ridge.

The available information is summarized in Table 3. The first column gives the reference number, the second the wave-length and the third the width of the antenna beam between half-power points. In the fourth column the half-width of the central source is given, measured above the background, in two co-ordinates, while aerial temperatures and flux densities are found in the fifth and sixth columns. The values of the flux densities are very uncertain, as it is difficult to decide which part of the central maximum should be considered the central source. A factor of two can easily be allowed for.

It is interesting to compare the 22 cm and 3·5 m results. At 3·5 m Mills[7] observes a minimum at the exact position of the 22 cm source. This minimum can be readily explained by assuming that the source at 22 cm consists of ionized hydrogen.

Correcting for beam-width we find a maximum temperature at 22 cm of roughly $300°$ K corresponding to an optical depth of 0·03. At 3·5 m the optical depth then would be about 6. Even at a distance of $1°$ from the centre, where the temperature is about $50°$ K, we have an optical depth of 0·005 at 22 cm and 1 at 3·5 m. From this we may conclude that over a range of $2°$ in longitude practically all radiation from behind the centre is blotted out. If the high-temperature maximum, observed at longer wave-lengths, is due to a central concentration of matter, radiating non-

31

Table 3. *Radio emission from the galactic centre*

Reference number	Wave-length in cm	Beam-width	Half-width of source	T_b (° K)	Flux density $\times 10^{24}$	Remarks
1	3	$0°13 \times 0°15$	$0°25$	?	4–5	Small source (no. 5) only
2	9·4	$0·45 \times 0·40$	Point	$17 (= T_a)$	4·8	No. 5 only. Broad source (no. 9) has $T_a = 6°$, $T_b = 10°$ K
3	21	0·92	?	45–$65 (= T_a)$	17	No. 5 + no. 9 (no. 9 has $T_b = 6°$–$30°$)
4	22	0·57	$0·83 \times 0·64$	166	35	Integration over nos. 5 and 9
5	33	3×6	Point	29	29	No. 5 + no. 9
6	50	3·3	Point	40	29	No. 5 + no. 9
7	75	2	Point	310	33	No. 5 + no. 9
8	75	2	Point	180	16·4	No. 5 + no. 9?
9	350	0·92	?	32,500	—	Source seen in absorption. 'Background' $5·5 \times 10^4$?
10	1500	1·4	2·4	200,000	—	Source seen in absorption. 'Background' 5×10^5?

References to the table

1. Haddock, F. T. and McCullough, T. P. *Astron. J.* **60**, 161, 1955.
2. Haddock, F. T., Mayer, C. H. and Sloanaker, R. M. *Nature*, **174**, 176, 1954.
3. Hagen, J. P., Lilley, A. E. and McClain, E. F. *Astroph. J.* **122**, 361, 1955.
4. Westerhout, G. *C.R.* **245**, 35, 1957.
5. Denisse, J. F., Leroux, E. and Steinberg, J. L. *C.R.* **240**, 278, 1955. See also reference no. 6.
6. Piddington, J. H. and Trent, G. H. *Austr. J. Phys.* **9**, 74, 1956.
7. Westerhout, G. *Bull. Astr. Inst. Netherl.* **13**, 105, 1956.
8. McGee, R. X., Slee, O. B. and Stanley, G. J. *Austr. J. Phys.* **8**, 347, 1955.
9. Mills, B. Y. *Observatory*, **76**, 65, 1956.
10. Shain, C. A. *Austr. J. Phys.* **10**, 195, 1957.

thermally, this might mean that the ionized hydrogen region is situated in the centre of the galactic system. The 32,500° K observed at 3·5 m in this direction is then partly due to foreground radiation (22,500° K) and partly to black-body radiation from the thermal source (10,000° K). In that case the source has blotted out 22,500° K from behind so that the peak temperature in the absence of the ionized hydrogen should have been 45,000° K.

Davies and Williams[8] and McClain[9] have observed the central source in absorption at 21 cm and conclude from their observations that it must be situated at 3 kpc from the sun. The initial data and the consequent interpretation are, however, very uncertain.

Substantial evidence for the source being the galactic nucleus itself has recently been obtained by van Woerden, Rougoor, and Oort[10]. In the direction of the centre they observed a maximum in 21 cm-line profiles, attributable to a small spiral arm, with a radial velocity of − 50 km/sec. It is seen as an absorption feature at the position of the centre source, indicating that it is lying in front of this source. If the source was situated at a distance of only 3 kpc from the sun, the small spiral arm would be lying in a region where thus far only small deviations from circular velocity have been observed. On the other hand, in the neighbourhood of the galactic centre, 21-cm line measurements show deviations from circular motion as large as 200 km/sec. They conclude that the source is situated in the central region and is probably to be identified with the nucleus of our Galaxy.

As regards the expansion observed in the nuclear region, Paranago[11] in a study of the K-effect has shown that some sub-systems of the Galaxy may be expanding while others are contracting. An alternative view was put forward by Lindblad. He suggested that perhaps there are some similarities between the galactic system and the barred spirals, where we have a very small nucleus from which threads of dark matter extend and where we very often find a small spiral in the centre of the system.

It is obvious that in the direction of the galactic centre optical measurements, particularly in the far infra-red, are urgently needed for comparison with the radio data.

References

[1] Gaposchkin, S. I. *Variable Stars*, **10**, 337, 1955.
[2] Oosterhoff, P. Th. *Bull. Astr. Inst. Netherl.* **13**, 67, 1956.
[3] Stebbins, J. and Whitford, A. E. *Astroph. J.* **106**, 235, 1947.
[4] Kaliniak, A. A., Krassovsky, V. I. and Nikonov, V. B. *Dok. Akad. Nauk S.S.S.R.* **66**, 25, 1949.
[5] Dufay, J., Bigay, J. H. and Berthier, P. *Vistas in Astronomy*, 1539, 1956 (Pergamon Press, London).
[6] Haddock, F. I., Mayer, C. H. and Sloanaker, R. M. *Astroph. J.* **119**, 456, 1954.
[7] Mills, B. Y. *Observatory*, **76**, 65, 1956.
[8] Davies, R. D. and Williams, D. R. W. *Nature*, **175**, 1079, 1955.
[9] McClain, E. F. *Astroph, J.* **122**, 376, 1955.
[10] Woerden, H. van, Rougoor, W. and Oort, J. H. *C.R.* **244**, 1691, 1957.
[11] Parenago, P. P. *Uspekhi Astr. Nauk*, **4**, 69, 1948.

(E) THE GALACTIC DISK

According to the general evolutionary ideas outlined in section (A), p. 13 above, we should find in the disk both old objects, formed immediately after the contraction of the primeval cloud into the disk, and recently formed stars. The study of the interarm population of our Galaxy is a difficult problem due to the interstellar absorption. Information from nearby galaxies, especially from M 31, is in some ways more readily obtainable.

In the disk of M 31 we find stars of absolute magnitude about -3 photovisual, which are similar to the brightest stars observed in globular clusters. Observations with the 200-inch telescope have shown that the number of these stars of population II decreases rapidly from the nucleus to the adjacent regions, and then slower towards the limit to which the system can be traced by optical means.

One of the most puzzling things concerning the disk of our Galaxy has been that in our neighbourhood stars of the sort found in the disk of M 31 could not be identified in any great amount. If the disk of the Galaxy contains population II we should expect to find in our vicinity many stars of the sub-giant type, for instance. Eggen's recent colour-magnitude diagram [1] for the stars in our neighbourhood shows a main sequence and a branching-off at about absolute magnitude 3·5. Both stars of the M 67 type and of the globular cluster variety occur around this branching point, but the further run of the branch indicates that we are dealing mainly with stars of the M 67 type. Nevertheless, as there is good reason to believe that the Galaxy is an Sb spiral like M 31, it is to be expected that an important component of the disk of our Galaxy should be objects of pure population II. At the conference, the following evidence in this direction was given by Baade.

(1) *Novae and planetary nebulae*

A general picture of the distribution of the disk population of the Galaxy would be obtained by means of objects of very high luminosity, like novae and planetary nebulae. The distribution of novae has been studied by Kukarkin [2] and by McLaughlin [3] who, using rough corrections for absorption, derived essentially the same picture. The novae are found to fill a disk-like volume. Only one nova, VY Aqr, situated at a distance of about 5 kpc from the galactic plane, appears to be associated with the halo. In M 31 the novae follow the general intensity distribution of the disk, showing no relation whatsoever to the spiral arms.

The same galactic distribution is shown by the planetary nebulae. They have somewhat lower luminosity than the novae, but in Hα light and objective-prism surveys they can be easily found even if apparently very faint. From the recent surveys by Minkowski [4] and by Haro [5] a rather complete picture of the distribution, including the galactic nuclear region, has been obtained. Haro remarked that several planetary nebulae of apparent magnitude 16 and fainter have been found in the polar caps; the distribution of the planetary nebulae may therefore be somewhat different from the novae in that a larger proportion are halo members. However, most of them belong to the disk.

34

(2) *Globular clusters*

When allowance is made for the absorption, the space distribution of the observed globular clusters appears to indicate an excess of objects at low galactic latitudes. A large number of these clusters must be completely hidden behind obscuration. Several investigators have tried to estimate the number of these; the figures vary from some 50 to nearly 100. We thus arrive at the conclusion that there is a disk-component of the globular cluster population.

Morgan [6] has shown that the integrated spectral classes of the globular clusters of the halo strongly depend upon the criterion used for the classification. He used a dispersion of 150 Å/mm and three criteria, the ratio of Hγ to the G band, the intensity of the hydrogen lines, and finally the intensity of the Fe I lines. According to the hydrogen intensities the halo clusters have spectral classes between F6 and F8, while slightly earlier spectral classes are obtained from the ratio of Hγ to G. The spectral classes, inferred from the iron lines, however, are 0·8 classes earlier. The globular clusters of the halo have weak metal lines compared with the strength of the hydrogen lines. From the luminosity function of the stars in globular clusters, as established by Sandage, we know that the integrated spectra refer to the stars of the three brightest magnitudes.

Morgan also studied a group of clusters to which Mayall had called attention, with considerably later spectral class. No discrepancy exists between the spectral classes inferred from the intensities of the hydrogen lines on the one hand and the metal lines on the other, and the clusters were found to be in the interval G2–G5. Morgan pointed out that their integrated spectra are similar to that of the nuclear region of M 31. The interesting point is that these late-type globular clusters have very small z-components in the Galaxy. The ten clusters of types G2–G5 in Mayall's list have a mean z-component of only 1·4 kpc. Thus it seems that we have here a spectroscopic method to pick out globular clusters of the galactic disk.

Very little is known about these disk clusters because many are situated in southern declinations and projected against star clouds. A study of two of them is being made by Sandage and collaborators (NGC 6356 and 6712). Cuffey is investigating the colour-magnitude diagram of NGC 6838 on plates taken at the Mt Wilson 100-inch. According to Mrs Hogg [7] only three of Mayall's G-type clusters have been searched for variable stars. In NGC 6838 four variables are listed, but apparently none of the stars is of the RR Lyrae type, in NGC 6356 five variables are listed and

35

in NGC 6712 twelve. According to Oosterhoff[8] one of the variables of NGC 6712, with the period of 105 days and amplitude about 1 magnitude, is similar to variables of this kind in other globular clusters. The search for variable stars in the G-type clusters should be pushed vigorously.

Of course not all globular clusters at low latitudes are to be regarded as disk members. Due to their motions across the galactic plane some of the halo clusters will be found within the disk at present. Such a case is NGC 6522, the centre of Baade's variable star field in Sagittarius. It has a distance of only 600 pc from the galactic plane, but, according to the spectral class, it is a halo cluster.

Baade advocated a search for new low-latitude clusters by means of the Palomar Sky Survey. These new clusters as well as previously known objects are then to be investigated by photo-electric and spectroscopic means. From this we may get a clearer picture of the distribution of the disk component of the globular clusters. One of the most difficult data to obtain in this study will be the intrinsic colours.

Thackeray mentioned in this connexion that within a year or so a rather large material on southern globular clusters will appear from the work of Kinman at Pretoria. Spectra and radial velocities, both for the integrated clusters and for individual stars, will become available.

(3) *RR Lyrae variables*

The first indication that some RR Lyrae stars belong to the galactic disk was found by Soviet astronomers[9]. This induced Struve[10] to rediscuss Joy's radial velocities of the brightest RR Lyrae stars. He found that the solar motion referred to the variables with periods shorter than 0·4 day was considerably smaller than that referred to variables with longer periods. The stars forming Joy's material are situated within 1500 pc from the sun. According to Baade, a plot of the number of variables at various distances from the galactic plane shows that even this material is very deficient due to absorption effects close to the galactic plane. At $z = 250$ pc the deficiency factor is about 5, and at $z = 0$ about 8. A set of radial velocities which is free of these deficiencies would undoubtedly show even stronger the admixture of low velocity stars among RR Lyrae variables with periods below 0·4 day. It is not clear yet how we can separate the disk members from the rest of the RR Lyrae stars. Perhaps a-type variables with periods below 0·4 day are to be sought for disk members, as is suggested by Oosterhoff's work[11].

By means of the RR Lyrae stars it may thus be possible to arrive at an estimate of the relative strength of the disk population II in our neighbourhood and the almost equally old population of the M 67 type. The problem is very difficult, as it includes the determination of colour excesses for the calculation of distances and z-values.

(4) *Long-period variables*

Long-period variables occur in different components of the Galaxy, which complicates studies based on these objects. On the other hand the long-period variables offer several advantages. The probability of their discovery from a comparison of plates of different epochs is so high that in the whole sky practically all stars of this type brighter than photographic magnitudes 11 or 12 are known. In numerous fields investigated at Harvard, Sonneberg, Leiden and Moscow all variables brighter than 15^m are known, and in addition we know at least all variables brighter than 18^m in a few smaller fields investigated by means of large telescopes. The periods and limits of light variation are easily determined even from small plate material provided a sufficient time interval is covered.

Some morphologic peculiarities of the long-period variables, in the first place the length of the periods and the form of the light curves, are definitely connected with the the kinematic characteristics. The majority of stars with periods between 160 and 220 days have symmetrical light curves and a large velocity dispersion and certainly belong to the halo. On the contrary, most stars with periods longer than 400 days have asymmetrical light curves and belong to the disk. Unfortunately, the separation of the population components by means of light curves is not quite definite, and it would be very valuable to have some spectroscopic criteria in addition. The spectroscopic differences seem to be very small, however. Keenan has recently made extensive studies at Mt Wilson to search for correlations between spectral peculiarities and light-curve characteristics, but the preliminary results show only the general spectrum-period dependence.

Kukarkin mentioned that his conclusions [12] of 1954 on the non-random distribution of the long-period variables in the polar caps are not substantiated, because certain selection effects had been neglected. He stressed, however, that the stars of this type show a tendency to group in the galactic star clouds. The periods of stars situated close to each other in space are usually also nearly equal.

Besides the difficulty of assigning specific variables as members of the halo, disk, etc. there are some other obstacles for using the long-period variables as an effective tool for galactic research. Most important is perhaps that only 400 of the 3500 known objects have been investigated adequately. Two years ago Soviet astronomers therefore started regular observations of these stars brighter than 12^m at maximum. Ashbrook at Harvard, Whitney in Oklahoma, Weber in Paris, and astronomers in Sonneberg, Stalinabad, Odessa, and Moscow are studying stars neglected up to now. Some results have already been obtained, the properties for more than 200 stars have been corrected, and certainly the characteristics of all long-period variables brighter than 12^m at maximum will be known within three or four years. It seems desirable to ask the amateur organizations to extend their programmes on these stars.

Another difficulty is our poor knowledge concerning the long-period variables in other galaxies. Baade stated that in M 31 the search for these variables on ordinary blue plates has led to only very few stars, but in the continuation of this work on photovisual plates the stars suddenly appear in large numbers. He hoped that especially from investigations of some of the E galaxies of the local group, for instance NGC 185, which has been studied ever since the 200-inch telescope came into operation, a rather complete picture of the distribution of long-period variables will be obtained. So far, Baade has intercompared five pairs of long-exposed photovisual plates of NGC 185 and found more than 500 red variables. From a preliminary study it is evident that there are several types among these. Predominant are periods around 200 days, but there are also a few about 340 days. Others, which are variable by about a magnitude or a magnitude and a half, have periods of about 100 days and are quite regular; they seem to be of the same type as those found in globular clusters. It was very difficult to accumulate this material, because only nights with the best seeing can be used. Baade has, however, collected about 100 plates since the study began, and he believes this material will provide answers to some of the questions concerning the long-period variables.

In the Magellanic Clouds we may have to go to about magnitude $17\cdot5$ on blue plates to find the long-period variables with periods around 200 days, and to magnitude 20 to find those around 400 days. This has not been done yet. Thackeray mentioned, however, that one long-period variable with a maximum around $17\cdot5$ has been found in the cluster NGC 121, which is situated in the Small Cloud.

Thackeray also called attention to the three long-period variables in 47 Tucanae. They have periods around 200 days and according to Feast their spectra are very similar to those of ordinary long-period variables. Parenago in this connexion raised the question if we are certain that there are no more than three long-period variables in 47 Tucanae, and if not more variables of this type could be found in this and other globular clusters. Thackeray said that there are some irregular variables present in 47 Tucanae, in addition to the three regular variables, but he doubted that there are more than these three regular ones present. The irregulars have M type spectra and at Pretoria several other stars in 47 Tucanae of spectral class M are known, which appear not to be variables. Baade expressed his great interest in Parenago's question and pointed out that Mrs Hogg's catalogue[7] contains several more clusters with which long-period variables seem to be associated. In all cases, however, the clusters are found projected against large Milky Way clouds and the variables are rather distant from the centres of the clusters. Baade stressed that the three long-period variables in 47 Tucanae seem to be a unique case.

Vyssotsky reported on recent work at the McCormick Observatory for determining the proper motions of long-period variables. These have been completed for 347 of the brighter variables, all but sixty-three of which are situated north of the equator. The proper motions are being reduced to the McCormick system, which is only slightly different from the FK 3 system. The radial velocities for 290 of these stars are known from the work of Merrill[13] and as stated by Thackeray, the radial velocities for some of the southern stars are being determined at Pretoria. An attempt will be then made to derive secular parallaxes for groups of stars, and hence absolute magnitudes and the space-motion distribution. Some difficulties will probably be met with in this connexion due to the non-uniformity of the observations, especially with regard to the apparent magnitudes, and the groups therefore have to be made rather large. A similar study was made several years ago by Ikaunieks[14].

(5) *Infra-red spectral surveys*

The infra-red spectral surveys carried out at the Warner and Swasey Observatory and reported by Nassau cover a belt 12° wide along the galactic equator from about longitude 200° through 0° to about 340°. Much of the observations and the discussions have already been published[15]. The disk-like distribution of the M stars, and the fact that these stars are greatly concentrated in the direction of the galactic nucleus, have been shown. All the BD stars of class M2 and later in the belt have now been classified, and lists have been published for M5 and later. Using the same plates some 700 new carbon stars and about seventy new S stars have been found. Lists giving approximate magnitudes for these stars have also been published. The carbon stars seem to be more concentrated to the galactic plane than the M stars, while the S stars show still greater concentration. In addition, all the red variables in the zone, as listed in the

General Catalogue of Variable Stars down to photographic magnitude 15, have been classified.

The current investigations deal with four fields in the Milky Way, namely, regions at longitude 33°, 41°, 55°, and 354° (the Scutum cloud). These are being studied by Westerlund, Velghe, McCarthy, and Albers, respectively. The studies include classifications of all stars between spectral types M2 and M10 down to 13th magnitude infra-red, and determinations of the mean interstellar absorption in each region.

In addition to this, two regions at higher latitudes are being studied, one at $l = 43°$ and extending up to $b = 22°$, the other at the longitude of the galactic centre and extending up to $b = 40°$. Some details concerning the work on the latter region have been given in a previous section of this report (p. 29).

References

[1] Eggen, O. J. *Astron. J.* **62**, 45, 1957.
[2] Kukarkin, B. V. *Astr. J. U.S.S.R.* **24**, 269, 1947.
[3] McLaughlin, D. B. *Astron. J.* **51**, 136. 1945.
[4] Minkowski, R. *Pub. Obs. Univ. Michigan*, **10**, 25, 1951.
[5] Haro, G. *Bol. Obs. Tonantzintla y Tacubaya*, no. 1, 1952.
[6] Morgan, W. W. *Pub. Astr. Soc. Pacif.* **68**, 509, 1956.
[7] Hogg, H. B. *Pub. David Dunlap Obs.* **2**, 33, 1955.
[8] Oosterhoff, P. Th. *Bull. Astr. Inst. Netherl.* **8**, 273, 1938.
[9] Kukarkin, B. V. *Investigations of the Structure and Evolution of the Galaxy based on the Study of Variable Stars* (Publ. House of Techn. and Theor. Lit., Moscow-Leningrad, 1949, in Russian; German translation Akademie-Verlag, Berlin, 1954).
[10] Struve, O. *Pub. Astr. Soc. Pacif.* **62**, 217, 1950.
[11] Oosterhoff, P. Th. *Trans. I.A.U.* **8**, 502, 1952.
[12] Kukarkin, B. V. *Astr. J. U.S.S.R.* **31**, 489, 1954.
[13] Merrill, P. W. *Astroph. J.* **94**, 171, 1941.
[14] Ikaunieks, J. J. *Variable Stars*, **8**, 393, 1952.
[15] See the series of papers by Nassau and associates in *Astroph. J.* **119**, 175, 1954; **120**, 118, 129, 464, 478, 1954; **122**, 177, 1955; **124**, 346 and 522, 1956; **125**, 195 and 408, 1957.

(F) SPIRAL STRUCTURE

Investigations of the structural form of the Galaxy are being actively carried on at a number of observatories. The portion of the Galaxy surveyed has been markedly increased since the time of the first Symposium for Co-ordination of Galactic Research. Many new results have been obtained, particularly in the southern hemisphere; additional observing programs are in the planning stage. New observational techniques for some problems have been found and are under investigation.

General survey of spiral structure as indicated by neutral hydrogen

In principle, the large-scale distribution of neutral hydrogen through-out the entire Galaxy can be observed in the 21-cm radiation of the neutral hydrogen atom. In practice, certain regions (sectors in the direction of the galactic center and anti-center) must be excluded because in them galactic rotation effects become very small or zero. Much useful information will be found in I.A.U. Symposium No. 4, *Radio Astronomy* [1].

In surveying the present status of 21-cm results on spiral structure, Westerhout first dealt at length with practical limitations on the accuracy attainable in the description of the hydrogen distribution. Antennas used for 21-cm spiral structure surveys generally have had small angular resolution. The Dutch 7·5-meter reflector, with which most of the surveys of the Milky Way have been made, had an elliptical antenna pattern of essentially gaussian form with half-widths of $2°78$ and $1°85$. A large volume of space was included in each pointing of the antenna; many hydrogen clouds were observed simultaneously. The 21-cm radiation received is spread in frequency by the differential galactic rotation of the clouds and by the random motion of the clouds. The observed line profile from a concentration of hydrogen clouds in a spiral arm generally has a broad hump with indications of small-scale internal structure.

To estimate the space density of hydrogen from such a profile, one must first statistically correct [2] the profile for random cloud velocities. Radio observations suggest a gaussian distribution of random velocities; optical observations [3] indicate an exponential distribution of velocities. The radio results may arise from an exponential distribution of velocities of the clouds plus a gaussian distribution originating in temperature broadening and turbulence in individual clouds. Little information on variation of the cloud velocity distribution in different parts of the Galaxy exists. For correction of the 21-cm line profiles it has been assumed that the velocity distribution is approximately gaussian with a dispersion of 15 km/sec at $R = 2$ kpc; the dispersion of random velocities is assumed to decrease to 6 km/sec for $R > 8$ kpc. Small changes in numerical parameters and in the reduction procedure produce large changes in the ratio of the space density of hydrogen in and between the arms. Inter-arm densities are particularly uncertain; they are especially sensitive to the corrections for random velocities.

Estimation of hydrogen density also requires that the gas temperature be known. In three galactic longitudes at distances within two or three kpc from the sun, large optical depths are reached; a harmonic mean temperature of 125° K has been found in these areas [4]. Lack of any additional information on gas temperature has necessitated adoption of the hypothesis that gas temperature is constant throughout space and of the value stated. This assumption must be regarded as a crude one. The temperature might well increase toward the galactic center; it might be higher in the arm than in the inter-arm region, or in the vicinity of groups of hot stars.

The rotation law of the Galaxy as a function of distance, R, from the galactic center must be known if a distance is to be assigned to a hydrogen

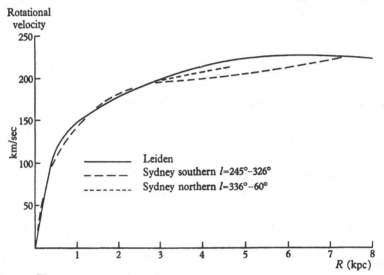

Fig. 1. Comparison of Sydney and Leiden galactic rotation curves.

cloud observed in a specified direction and moving with a measured radial velocity. In the inner portion of the Galaxy the rotation curve can, with some accuracy, be established from the 21-cm observations themselves. The agreement of the rotation curve derived from observation in the ranges $l = 245°$ to 326° at Sydney [5] and $l = 336°$ to 60° at Leiden [6] is only fair. The distance scale is known only with the same accuracy as the distance to the galactic center. Fig. 1 shows the agreement of the northern and southern observations. On the basis of a rotation curve drawn through the Leiden observations (fig. 1), and utilizing optical data of various kinds, Schmidt [7] derived the mass distribution in the Galaxy. From this mass model the rotation curve for the outer part of the Galaxy was derived; circular motion was assumed throughout. Large uncertainties in the numerical values of the parameters A, B, R_0, V_0, and the mass in model

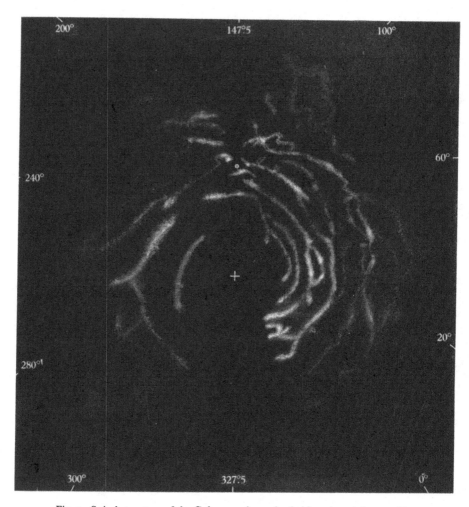

Fig. 2. Spiral structure of the Galaxy as shown by Leiden observations and by Sydney observations.

facing p. 43

produce large uncertainties in the distances derived for the hydrogen clouds.

The reduction of the 21-cm observations is thus subject to numerous uncertainties; details of the derived hydrogen distribution are unreliable. However, the broad features of the picture of spiral structure resulting from these observations are well established. Results obtained from overlapping observations at Leiden [8] and Sydney [9] show very satisfactory agreement. Fig. 2 illustrates the picture of spiral structure derived from the combined Leiden and Sydney observations.

Lindblad [10] spoke about the hypothesis of circular orbits utilized in the interpretation of the 21-cm observations. He drew attention to what he termed 'dispersion orbits' in the Galaxy. These are defined as the orbits that a cloud of free particles—the members of a stellar association, for example—pursue as the group is disrupted by differential galactic rotation. He expressed the belief that a close relation must exist between such orbits and spiral arms. If κ represents the frequency of oscillation along the radius vector in a galactic orbit that differs little from a circle, Lindblad asserts that at those R values for which $n = \dfrac{d\kappa}{d\omega} = 1$ or 2 the dispersion orbits are closed. For other values of n the orbits are more complicated figures and are not closed. At those R values where closed dispersion orbits occur, permanent, slightly elongated rings of material are to be expected. These concentrations of matter Lindblad identifies with spiral arms. From the galactic rotation curve derived at Leiden, Lindblad has found $n = 2$ at $R = 4\cdot3$, $4\cdot7$, $5\cdot3$, $6\cdot5$ and $8\cdot3$ kpc, and $n = 1$ at $10\cdot5$ and 13 kpc. These values are closely similar to those at which the Leiden observers find concentrations of hydrogen.

The most probable shape of a dispersion orbit is circular, though in nature, deviations may be expected. Fig. 3 a shows a group of hypothetical dispersion orbits suggested by the picture of spiral structure in the Galaxy. The true arms of the Galaxy are represented by the concentration of material in these orbits. Motion in these orbits is not purely circular; there is a radial component as well. The precise motion in the orbit is known from the properties of the galactic model adopted for the calculation. Fig. 3 b displays the form of the spiral arms that would be deducted if radial velocities of matter in the arms (Fig. 3 a) were observed from the sun and if distances were derived for that matter under the assumption of circular motion. Great distortion of the true picture (Fig. 3 a) results from such treatment. Lindblad drew attention to the fact that the structure pictured in Fig. 3 b bears some resemblance to the picture of spiral arms derived from the 21-cm observations (fig. 2). The conclusion to be drawn is clear; the possibility of an alternative interpretation of the arm pattern on the basis of non-circular motion should not be overlooked.

Considerable discussion revolved around the question of how to test empirically the commonly accepted hypotheses of circular motion.

Lindblad suggested a simple test might be made by looking for velocity deviations in the center and anti-center regions, where, on the hypothesis of circular motion, the velocity should be zero. Westerhout reported that in the anti-center regions, local velocity deviations up to 30 km/sec are found in the hydrogen clouds. In the direction of the galactic center a concentra-

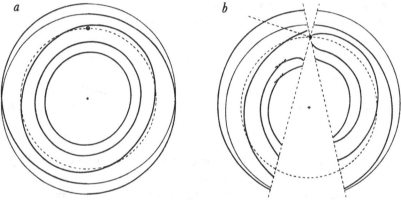

Fig. 3a. Dispersion orbits.
Fig. 3b. 'Apparent' structure for the same orbits shown in Fig. 3a deduced from radial velocities relative to the circular motion at the sun when these are interpreted as due to circular motion.

tion of hydrogen approximately 2° wide and more than 15° long is observed to be moving radially away from the galactic center. Further work on the question of velocity deviations is highly desirable.

An indirect approach to the question of circular motion might be made by testing the similarity of (a) the form of spiral structure derived by using radial velocities and the circular orbit hypotheses to determine distances, and (b) the form of spiral structure derived by some other completely independent method.

The hydrogen layer appears to be of rather constant thickness. Kerr proposed that the forms of the hydrogen spiral arms be based on distances derived from high resolution measurements of their angular thickness. Oort was not optimistic about the successful application of the method. Efforts at Leiden to separate nearby and distant spiral arms on the basis of this principle were not entirely satisfactory. Westerhout pointed out that in the outer parts of the Galaxy the thickness of the hydrogen layer appeared less constant than in the inner parts of the Galaxy. Here it would appear well to point out, however, that present estimates of the thickness of the hydrogen layer depend upon the velocity model and involve the assumption of circular motion.

According to Oort a satisfactory test for non-circular motion could be made by study of the galactic clusters. Photometric distances of the clusters could be found. Radial velocities could be measured and departures from a motion field circular about the galactic centre could be determined. A rather large number of clusters would be needed. The calibration of the absolute magnitudes and intrinsic colours to be used in the photometric distance scale is an important part of this approach to the problem.

Other galactic parameters, R_0 and V_0, also appear in the calculation of distances from galactic rotation effects. Weaver[11] stated that the values of V_0 presently derived by different methods are inconsistent. Investigation of the cause of the inconsistency is needed. The product $V_0 = R_0 \omega_0$ with $R_0 = 8 \cdot 2$ kpc and $\omega_0 = 26 \cdot 4$ gives 216 km/sec. The estimate of V_0 derived from the solar motion determined by Humason and Wahlquist from the local group of extra-galactic nebulae is 260 km/sec. The value includes the component of peculiar motion of the Galaxy in the direction of rotation; however, it indicated $V_0 > 200$ km/sec. Fricke used the high-velocity stars to calibrate the Bottlinger diagram. He found $V_0 = 276$ km/sec. Mayall used radial velocities of 50 globular clusters to derive $V_0 = 200$ km/sec. This fundamental determination underestimates V_0 by the galactic rotation of the system of globular clusters. Schmidt estimates that the average rotational velocity of the globular cluster system is 80 km/sec. The product $R_0 \omega_0$ appears to underestimate V_0. Arguments can be given to show that ω_0 cannot be far wrong; it is likely that $R_0 = 8 \cdot 2$ kpc is an underestimate. Even if motion is circular in the Galaxy, errors in the distance scale caused by an error in R_0 will seriously distort the picture of spiral structure. A better determination of the numerical value of R_0 is greatly needed.

Oort remarked that little weight can be attached to an estimate of V_0 derived from members of the local group of nebulae. The peculiar motion of the Galaxy is so large that it vitiates the result.

Tests of the circularity of galactic motions, the determination of spiral arms by methods independent of galactic rotation, and the galactic distance scale remain problems of high priority in galactic studies. Much work needs to be done on the random velocity distribution of interstellar material, and on the problem of variation of the temperature of neutral hydrogen throughout the Galaxy.

The z-distribution of neutral hydrogen

Kerr, Hindman, and Carpenter[12] have utilized observations of the 21-cm radiation of neutral hydrogen to investigate the z-distribution of the hydrogen. They found the gas layer to be thin, with an approximately constant thickness of 200–250 pc between half-density points. In the inner portion of the Galaxy the surface formed by all maximum density points is very flat. Deviations from an average plane (the 'principal plane' of the Galaxy) are of the order ± 35 pc. In the outer parts of the Galaxy, the surface formed by the points of maximum density is systematically distorted[13]. The effect is large scale, with little relation to the spiral pattern. Maximum distortion occurs in the direction toward and away from the Large Magellanic Cloud. The effect suggests a gravitational tide, but is much too large for a simple gravitational explanation of this nature if the mass of the Cloud is of the order of magnitude generally assumed. The effect may involve forces other than gravitation; it may be an after-effect of some earlier event in the history of the Galaxy. No decisions on such questions now seem possible.

It is of interest to note that the mass of M 32, estimated under the assumption that it gravitationally produces the distortion in M 31, is also unexpectedly large[14]. Other systems provide evidence that there are large interactions between galaxies. NGC 4565, an edge-on spiral, shows edge distortion similar to that observed in the Galaxy.

Oort expressed the opinion that the deviations in the z-direction observed in the outer portion of the Galaxy represent the remnant of some original structural feature of the Galaxy. The idea that they are caused by the Magellanic Clouds is not attractive. Blaauw questioned whether the strong departures from a plane in the outer parts of the Galaxy might not be in those regions where the period of oscillation in z is equal to the rotation period. In the inner region of the Galaxy the two periods may differ. The suggestion is an interesting one. However, the equality of period pictured might be the case only at rather large R values. Kerr remarked that the symmetry stretches over a rather long range in R and therefore the explanation involving the Magellanic Clouds appeared rather plausible to the Sydney group.

The galactic pole determined from observations of neutral hydrogen

The highly flattened hydrogen sub-system in the inner portion of the Galaxy provides the data for an accurate determination of the galactic pole. Two determinations have been made from the 21-cm observations.

At Leiden [15] a solution has been made on the basis of observations of the whole region inside the sun's circle on the northern side of the galactic center. The Sydney solution is derived from observations of the 'tangential points' on both sides of the galactic center. The sources of uncertainty differ in the two solutions shown in Table 4, but the derived values are in satisfactory agreement. Results very similar to those in Table 4 have been derived from the Cepheid variable stars [16] and from recent high resolution surveys of the radio continuum.

Table 4. *Position of the galactic pole derived from 21-cm observations*

	Galactic longitude	Galactic latitude	Sun's distance from galactic plane (in pc)
Leiden	$332°\pm8°$ (p.e.)	$88°63\pm0°07$ (p.e.)	$+23\pm10$ (p.e.)
Sydney (preliminary)	$349°\pm5°$	$88°34\pm0°25$	0 ± 32

The 21-cm radiation appears to fix the position of the galactic pole with high precision. However, the hydrogen represents a small percentage of the mass of the Galaxy. Blaauw called attention to the desirability of deriving the position of the pole from a very different kind of population— a disk population, but one that is not young, and that is likely to be symmetrically distributed. Objects found in the Warner and Swasey Observatory infra-red surveys [17] might be suitable for this purpose. It was agreed that a study of the distribution of such objects would provide a valuable check on the position of the pole derived from hydrogen. But it will be difficult to establish the pole with adequate accuracy, one- or two-tenths of a degree, from study of such a population.

Parenago pointed out that a powerful condition could be imposed on any solution for the pole if the direction of the galactic center could be fixed from radio or infra-red observations. However, it would be difficult, it was agreed, to fix the direction to the galactic center with high precision.

(2) *Continuous radio radiation*

Oort reviewed the present observational status of the continuous radio radiation as related to structure in the plane of the Galaxy. Three types of such radiation can be distinguished [18].

1. Thermal radiation arising from H II regions.

2. Non-thermal radiation originating in localized radio sources.

3. Non-thermal radiation probably originating in a continuous medium; source unknown, but possibly high energy electrons in interstellar magnetic fields.

A survey of the thermal component of the continuous radiation provides knowledge of the distribution of ionized hydrogen throughout the Galaxy. However, when observations are made, both the thermal and non-thermal components of radiation are received. Separation of the two components is difficult though, in principle, possible. The thermal component of the radiation varies as ν^{-2}; the non-thermal component varies approximately as $\nu^{-2\cdot6}$. In the non-thermal case the numerical value of the exponent is not accurately known. It varies somewhat with ν. It has been evaluated from studies of discrete sources and from studies of continuous radiation at high galactic latitudes.

It has been found empirically that, at least in the galactic plane, the thermal and non-thermal components of the radiation are of comparable

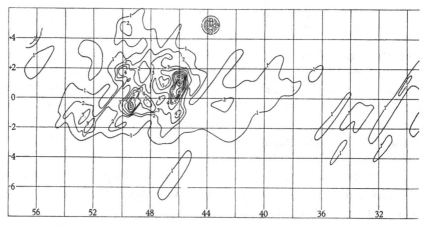

Fig. 4. Intensity-distribution of 22-cm radiation in the Cygnus region.

intensity at a wave-length of 20 cm. Observations at wave-lengths of 20 cm or less should thus show the galactic distribution of ionized hydrogen. Westerhout has utilized the 25-meter radio-telescope at Dwingeloo to complete a 22-cm survey in the latitude range $+6°$ to $-8°$ extending from longitude 320° through 0° to 56°. The half-width of the antenna pattern is of the order one-half degree. An example of the distribution found is shown in the map of the Cygnus region shown in Fig. 4.

A large part of the 22-cm radiation can be explained as thermal radiation from known H II regions. However, optical data are not sufficiently complete to permit a thorough comparison of optical and radio data.

Radio surveys at still shorter wave-lengths are required. At these shorter wave-lengths resolution becomes higher; the relative strength of the thermal component becomes greater. Individual H II regions should be identi-

48

fiable over a large part of the Galaxy. The thickness of the ionized hydrogen layer in the Galaxy should be well determined.

At present the most complete radio survey of continuous radiation is that at a wave-length of 50 cm made by Piddington and Trent[19] in Sydney. A large part of the radiation at 50 cm is non-thermal in origin. This is clearly indicated by the fact that, in a direction inclined at a slight angle to the galactic plane, the optical data (which provide a measure of the strength of the thermal component) indicate much less radiation than is actually observed by Piddington and Trent.

Basically, the results of the 50-cm survey resemble those of the 22-cm survey. The picture of spiral structure is well reproduced in both surveys. Since only approximately one-third of the observed 50-cm radiation is thermal in origin, we must conclude that we are here observing non-thermal radiation connected with the spiral arms. Kerr indicated that the spiral-structure pattern is shown in the $3\frac{1}{2}$-meter survey made in Sydney by Mills. This observation strongly supports the hypothesis of non-thermal radiation in the arms.

The non-thermal radiation in the arms could be produced by cosmic-ray electrons moving in interstellar fields. It is probable that such fields are stronger in the arms than outside the arms. As surveys at longer wave-lengths become available, it will be of particular interest to compare in detail the distribution of thermal and non-thermal radiation in the vicinity of the arms, to determine whether the layer in which the non-thermal emission originates may be thicker than the layer from which the thermal radiation comes.

In longer wave-lengths the galactic halo shows strongly. Recent observations[20] indicate that neutral hydrogen clouds of quite high velocity exist in the halo. These clouds are probably far from the galactic plane.

In regard to the third component of continuous radio radiation, the galactic radio sources, little can be said. A number of galactic sources have been identified, but no systematic survey has yet been made. A complete survey of such sources that will yield information on the general physical characteristics and the galactic distribution of such sources remains an important future problem.

Westerhout gave additional information on the occurrence of neutral hydrogen in the halo. Secondary maxima in the 21-cm profiles interpreted as these clouds indicate very low antenna temperatures, of the order of one to one and one-half degrees. They indicate predominantly negative velocities of approximately 40 km/sec, with a maximum velocity of 200 km/sec. The reality of such clouds in the halo needs careful checking.

Considerable discussion centred on the question of whether the 22-cm survey, say, should be used as a guide in the search for the largest associations and H II regions, or whether such a search should be made completely independently. Oort expressed the view that a survey at 10 cm would be preferable for a guide in such an optical survey. Several radio installations will be capable of making such a survey in the future. Westerhout and Blaauw felt that the 22-cm survey could serve some useful purpose as a guide in an optical search. Westerhout and Kerr indicated that at 22-cm wave-length the galactic ridge does not show over the longitude range approximately 60° through the anti-centre to 260°. A 10-cm survey would probably be even more confined to the region of the galactic center.

(3) OB star surveys

Morgan discussed two main topics: (a) the importance of continued and expanded observations of O associations; (b) means of discovering such associations over great distances.

Morgan pointed out that O associations contain very bright supergiants, objects five to ten magnitudes brighter than exist in clusters such as the Pleiades and Hyades. The statement was illustrated by an H-R diagram[21] of the I Sco association, which has NGC 6231 as its nucleus. It contains many super-giants earlier than Bo. Such stellar groups are of great interest from an evolutionary point of view; they also provide a means of delineating spiral structure to very much greater distances than is possible through observance of ordinary galactic clusters. Evolutionary differences between associations will complicate the calibration of the absolute magnitudes of the association stars. However, Morgan believes calibration of the required accuracy can be achieved.

Morgan described several methods for finding O associations:

(a) O associations can be discovered in the course of H II surveys[22]. Not all H II regions have their origin in O associations. But those H II regions that are connected with O associations have a characteristic appearance: they almost always appear to have the form of an incomplete circle. The method of finding O associations by the form of their related H II regions fails when the general star field is extremely rich. Even if one employs an interference filter, it is difficult to delineate precisely the regions of emission. Here Courtès' Fabry-Perot method[23] in which interference fringes are superimposed over an unresolved stellar background might prove useful. Not all associations contain stars early enough to ionize the gas in their neighbourhood. These would not be discovered by the H II-region technique.

(*b*) Three-color photometry [24] in wave-length ranges corresponding roughly to U, B, V provides a means of separating super-giants from other stars. Preferably a Schmidt telescope should be used. The three images are taken on one plate; the exposures are of such duration that the different intensities in the ultra-violet of different stars are clearly brought out. The three-color method makes use of the fact that an early-type star, even if reddened, produces a stronger ultra-violet image than an intrinsically red star.

(*c*) An alternative method [25] based essentially on the same principle as the three-color procedure makes use of very small dispersion spectra of 10,000–30,000 Å/mm. Inspection of the energy distribution in these very short spectra permits early-type stars (whether reddened or unreddened) to be separated from intrinsically red stars.

Sandage called attention to the importance of the O associations in the extra-galactic distance-scale problem. Distances of extra-galactic nebulae are derived from the brightest stars in the nebulae. Possibly the most important method of determining the absolute magnitudes of such stars is through study of associations in our own Galaxy.

One aspect of associations displayed in extra-galactic systems so far has not been observed in the Galaxy: super-associations. By this term Baade, who introduced the topic, referred to such objects as the star cloud NGC 206 in the south-preceding arm of the Andromeda nebula. This object, of length about 800 pc, appears to have A stars for its brightest members. In IC 1613 an irregular member of the local group, all blue stars are concentrated in one section of the system. Search for (and subsequent study of) such super-associations in our own Galaxy would be valuable. Heckmann reported that the big Schmidt at Bergedorf was being used for a general spectral survey. The prism, of ultra-violet glass, produces a dispersion of 570 Å/mm at Hγ. The limiting magnitude of the survey is approximately 13·5. At present, only high luminosity stars are being noted. The complete catalog, which will contain co-ordinates and charts, will probably include more than 10,000 objects. From the work done so far, it appears that the O and B stars are not entirely crowded together in small groups; there are OB stars between the groups. Eventually, the program will be extended to fainter stars with the aid of a second prism producing approximately half the dispersion used for this first survey. This second prism will be used at both Bergedorf (with the big Schmidt) and at Bloemfontein (with the ADH Schmidt).

(4) *Cepheid variables and galactic structure*

The classical Cepheid variable stars provide an important source of information on the form and kinematic properties of the Galaxy. However, many of the numerical parameters describing basic physical properties of the Cepheids require improvement before the full usefulness of these stars can be realized. In particular, the zero-point of the period-luminosity relation and the intrinsic colors of the Cepheids are not known with sufficient accuracy. The intrinsic dispersion in the period-luminosity relation is unknown. Even more basically, the question of the uniqueness of the relation needs investigation.

Oosterhoff described the observations and results of the Leiden Southern Station photo-electric program on Cepheids carried on with the Rockefeller Astrograph in the period 1953–56 by Walraven, Muller and Oosterhoff. In all, 182 Cepheids were observed in two colors (yellow and blue). These were mainly southern stars; a few northern Cepheids were included for comparison purposes.

Observations were confined to maximum light except in a few instances in which ephemerides were badly in error. A few standard stars were observed to permit reduction of the instrumental photometric system to the Cape 1953 S system[26]. To assign absolute magnitudes, Oosterhoff employed the period-luminosity relation determined by Shapley[27], with zero-point corrected by $-1 \cdot 7$ magnitudes. This correction, derived from seventeen stars, was applied rather than the value $-1 \cdot 4$ or $-1 \cdot 5$ magnitudes found earlier since previous investigators[28] had made insufficient corrections for interstellar extinction. For intrinsic colors Oosterhoff assumed (*a*) that galactic and Magellanic Cloud Cepheids are identical in properties; (*b*) that, as found by Code[29] and by Feast[30], the spectral types of long-period Cepheids at maximum light range from F5 to early G. He adopted the relation $\text{c.i.}_{\text{max}} = 0 \cdot 01 + 0 \cdot 10 \log P$. For the ratio of total photographic extinction to color excess Oosterhoff adopted the value $3 \cdot 5$, derived from a discussion of published values[31]. The Cepheids were observed at maximum; the period-luminosity relation refers to median magnitude. The observations were corrected to median magnitude by means of a statistical amplitude-period relation: $\text{amp}_{\text{pg}} = 0 \cdot 32 + 0 \cdot 91 \log P$.

Fig. 5 shows the galactic-plane projection of the spatial distribution of population I Cepheids as derived from the Leiden Southern Station observations and the fundamental data described above. Several stars considered to be population II Cepheids on the basis of *z*-height have been omitted. The average *z*-value of the population illustrated in

52

Fig. 5 is -23.9 ± 5.5 (m.e.) pc; the z-dispersion of the same population is 65 pc.

Interstellar extinction was found to be very irregular over the sky. All stars together produce a negative correlation between extinction per kpc and distance. In the Carina region there is no correlation between extinction and distance; in the Sagittarius region the correlation is positive. In the directions in which extinction is small, one sees to great distances; where it is large in relatively nearby regions, one sees to small distances. This situation results in the negative correlation observed for all stars.

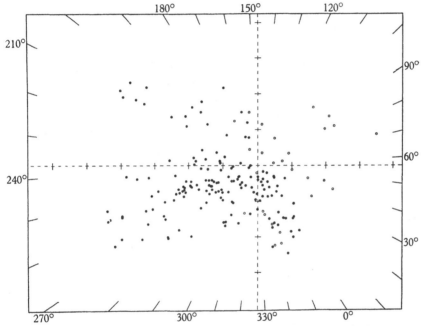

Fig. 5. Distribution of Cepheid variable stars in the galactic plane.

From fifty Cepheids for which radial velocities[32] were available, Oosterhoff found the value of the Oort A-parameter to be 17.4 ± 2.1 (m.e.) km/sec kpc.

In Fig. 5 the OB associations observed by Morgan, Whitford, and Code[33] are indicated by open circles; Cepheids by filled circles. The most distinct feature of the diagram, Oosterhoff said, is the concentration of Cepheids in Carina. This concentration indicates, he stated, that the line of sight falls in the Carina arm for a distance of three or four kpc. He identified this group of stars with the neutral hydrogen concentration shown in Fig. 2. Oosterhoff indicated that the Carina arm delineated by the

Cepheids can be followed to the neighborhood of the sun. Many of the Cepheids observed are in the direction of Sagittarius, but they do not belong to the Sagittarius arm. The Leiden Station observations did not extend beyond $l = 20°$, hence Oosterhoff could not trace the Carina arm farther. He suggested that the associations I Vul and I, II, III Cyg form a continuation of this arm. Oosterhoff assigned to the Sagittarius arm what he termed a second concentration of Cepheids lying in the longitude-range 200° to 10° and at a distance of $1 \cdot 5$–2 kpc. He called attention to the existence of seven OB associations involved with this more distant group of Cepheids. Only one association is found in the nearer Carina arm. A number of Cepheids are found in longitude 236°; the Sydney group finds no hydrogen concentration here. Oosterhoff suggested that these Cepheids may belong to the Orion arm.

The distance limit to which Cepheids are observed is very irregular with longitude. Many Cepheids must remain to be discovered.

The importance attributed to the Cepheids as galactic probes was indicated by the extent of the discussion that followed Oosterhoff's communication. Blaauw made the important point that the precision with which the population I Cepheids delineated spiral arms might depend upon the periods of the variables investigated. If period is an indicator of the evolutionary stage of the Cepheid (shorter periods being associated with older stars) one would expect the longer-period objects to define the arms more precisely than the short-period ones. Tests of this hypothesis should be made in the Galaxy and in members of the local group of galaxies. The investigation could be planned on a rather broad base. The distribution of periods in concentrations of Cepheids might be correlated with the type of environment in which the concentration is embedded. The characteristics of the environment here serve as indicators of probable evolutionary stage.

Morgan outlined a test for the relatedness of groups of Cepheids and OB stars presumably coincident in space: such groups should have the same average radial velocity. This direct test could be applied to many of the concentrations delineated by Oosterhoff. It should find frequent application in future investigations.

Discussion of photometric aspects of the Cepheid observations emphasized several points. It is imperative to investigate in detail the photometric system on which the observations are made. For a *general* program on Cepheids (or any other type of star) a standard photometric system, preferably U-B-V, should be adopted. Careful choice of filters or, for photographic work, filters and plates, enables one to reproduce very closely the response functions of the standard system. Transformations, if required at all, are then very small and the customarily employed linear transformation formulae are adequate for the purpose. Quantities well-established on the standard system, as, for example, the ratio of total photographic extinction to colour excess, can be used directly or with a very small transfer correction. Transformation uncertainties are not unnecessarily introduced in the final results.

Of special importance is the problem of the intrinsic colours of the Cepheids. Intrinsic colors must be known if an accurate zero-point of the period-luminosity relation is to be derived; they must be known if the period-luminosity relation is to be used in conjunction with observed magnitudes and colors in the determination of accurate distances of the Cepheids. In this latter connection, an error of one-tenth magnitude in an intrinsic colour introduces an error of approximately 12 % in the distance of the star. That the Cepheids are intrinsically bluer than was formerly believed has been established (a) from studies of Cepheids in the Magellanic Clouds [34], (b) from investigations of Cepheids in fields in which color excess as a function of distance modulus has been established by study of early-type stars or from a model of the reddening medium [35], (c) by observing polarization of the light of Cepheids previously supposed free from interstellar extinction effects [36], (d) by identifying the Cepheids, through spectral resemblances, with the non-variable stars of type I b. No one of these methods of investigation permits derivation of intrinsic colors free from all objections. It is not certain that Magellanic Cloud Cepheids are identical with galactic Cepheids or that the Clouds are free from interstellar extinction, that mean color excesses, determined as functions of distance in a particular region of the sky, apply in the particular direction of the Cepheid, or that the color-excess function is not a step function rather than a uniform function as is customarily assumed; that the relation between extinction and polarization is any more than statistical; that the variable Cepheids do indeed resemble the non-variable I b stars.

Kraft [37] and van den Bergh [38] have pointed the way to one solution to the problem of the intrinsic colors and absolute magnitudes of the classic Cepheid variables that may be free from these objections. The colors and absolute magnitudes may be found by investigation of Cepheids that appear to be probable members of galactic clusters. They have prepared lists of twelve clusters containing thirteen Cepheids with periods from 3 to 10 days. These stars should serve as fundamental checks on zero-point corrections and intrinsic colors established by other methods.

Considerable interest attaches to the size of the cosmic scatter in the period-luminosity relation. Knowledge of the dispersion about the period-luminosity relation is of importance in the evolutionary picture of Cepheids; it provides an estimate of the accuracy to be expected for a distance determination made through use of the relation. It is of interest to try to reduce the scatter about the relation by investigating light curves and other characteristics of the Cepheids and correlating those characteristics with departures from the present period-luminosity relation.

Kukarkin reviewed the present observational situation of Cepheids and spoke about problems requiring investigation. More than 600 Cepheids have been discovered in the Galaxy. Eggen[39] has photo-electrically observed thirty-two northern Cepheids in two colors; Oosterhoff has obtained two-color photo-electric results for 182 southern Cepheids at maximum light; Irwin[40] has two-color observations for 140 southern Cepheids; Mianes, at St Michel, is making two-color photo-electric observations of all Cepheids in the northern sky brighter than magnitude 13 at maximum. The spectrographic observations of Joy[41] and of Stibbs[42] have furnished radial velocities for approximately 200 Cepheids.

We know very little about very distant Cepheids in the Galaxy. It is important that large telescopes be employed to search for faint Cepheids in at least ten fields in transparent regions along the galactic equator. Very many Cepheids remain to be found. On the basis of the number of Cepheids in M 31, Baade estimates that there are 30,000 Cepheids in our own Galaxy. Rosino at Asiago is presently investigating three fields in Cygnus, Cassiopeia and Monoceros for faint variable stars. Additional investigations, especially in the southern hemisphere, are needed. The small z-dispersion of the classical Cepheids means that one could confine a search for these stars to a very narrow zone of galactic latitude not more than a few degrees wide. Population II Cepheids will be more difficult to find; information on these objects is needed also.

An accurate complete morphological study of Cepheids in the different components of the Galaxy would be of great value. Properties of Cepheids in the spherical component differ from those of Cepheids in the disk. The variety of differences appears to be large and, at present, poorly understood. Investigations of period changes deserve special attention. Such changes can be established with great accuracy; they appear to be of different character in Cepheids belonging to different galactic components. Parenago reported that from an investigation of forty or forty-five Cepheids he found period changes to be essentially discontinuous, and not proportional to time. Rate of change of period appears to be greater for long-period Cepheids, and is greater for Cepheids in the spherical component of the Galaxy than for Cepheids in the disk component. The changes observed are small. To detect them, one requires a long series of observations. Physically, the changes must correspond to variations in radius or absolute magnitude or both. Absolute magnitude changes of one- or two-thousandths of a magnitude might be expected; comparably small changes in radius might also be expected. Observatories with large plate collections of Cepheids could make a valuable contribution by studying period changes. Kukarkin has lists and maps of northern Cepheids. These are available to any observatory wishing to work on the Cepheid problem.

A general program of accurate observation of Cepheids now known and as they are newly discovered is work of the greatest importance for a more complete understanding of the Galaxy.

(5) *Spectrophotometric-photometric investigations*

Ramberg reported that he was engaged in spectrophotometric-photometric studies of several regions in and near the galactic plane in longitudes related to special directions in the Galaxy (in the direction of rotation, opposite to the direction of rotation, and so forth) in order to obtain information on large-scale galactic structure. The specific fields examined were chosen on the basis of homogeneity of a real star density and freedom from obscuration. He reported in detail on observations of 3063 stars in an area of 11 square degrees in Lacerta ($l = 69°$, $b = -8°$). The observations were made with the 40-cm Stockholm astrograph; the limiting magnitude is 13·25. Photographic and photovisual magnitudes were derived on the International System. Unwidened spectra were traced on a microphotometer and classified in a rough two-dimensional system.

The discussion of galactic structure was based on diagrams of measured color index plotted as a function of apparent photographic magnitude for individual spectral types. For stars of spectral type F0 to F8 and dG0 to dK0 the color index as a function of apparent magnitude shows no trend up or down. Ramberg interprets this to mean that the region of observation is very clear to a distance of 900 pc, at which distance he estimates a total interstellar extinction of 0·16 magnitude photographic. For the limiting magnitude of the survey, the dwarf stars do not permit observations beyond 900 pc. Plots of color index as a function of apparent magnitude for the giant stars G0 to K6 show, at a certain apparent magnitude, a sharp increase in apparent color. The fact that all classes show the increase at essentially the same apparent magnitude is interpreted to mean that the stars all have similar absolute magnitudes. At the same point at which the sharp rise in color occurs, an increase in number of stars takes place. Ramberg interprets this to mean that at a distance of approximately 900 pc the cone of view enters a concentration of stars and dark material.

The A-type stars are of special interest. The sharp increase in color sets in at different apparent magnitudes for the A3, A2, A0 stars. The differences between these values and the value found for the G-K giants indicate differences in the absolute magnitudes of the various spectral types. For the A0 stars the sharp rise in color starts at $m_{pg} = 10\cdot00$, increases to a maximum at $11\cdot80$ and thereafter remains constant. This latter feature signifies that the line of sight finally passes through the obscuring cloud and reaches a second relatively clear region.

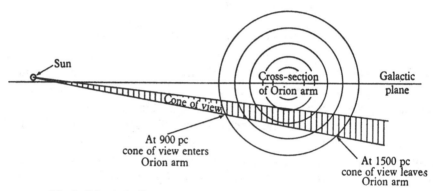

Fig. 6. Schematic diagram showing interpretation of the observed stellar distribution in Lacerta.

The physical interpretation of the observations given by Ramberg is shown schematically in Fig. 6.

The line of sight crosses the galactic plane and goes through the Orion arm. In the region of the arm Ramberg finds that the space density of A-stars increases by a factor of about two over what it is observed to be before entering the arm; the space density of the late-type giants increases by a larger factor. This would seem appropriate, since the giants are less concentrated to the galactic plane than the A-stars and the line of sight crosses the arm at an appreciable distance (about 170 pc) below the plane. There may be some indication of a neutral hydrogen concentration in the same general region as the region of increased star density observed by Ramberg.

A few faint B stars were observed in the Lacerta field. Ramberg finds these to be at a distance of approximately 5000 pc, probably in the Perseus arm.

58

(6) *Wilson's method of determining absolute magnitudes of late stars*

Wilson and Bappu[43] have made an extensive test of Wilson's[44] method of estimating absolute magnitudes of late-type stars (G through M) by measuring the widths of the Ca II emission lines. Coudé spectrograms of dispersion 10 Å/mm were employed. Whenever possible, both H and K emission lines were measured. The wave-lengths of the shortward component and of the longward component, determined with respect to metallic absorption lines in the stellar spectrum, provide data for determination of the line width W, expressed in km/sec. The line width W, when corrected for instrumental broadening by subtraction of 15 km/sec, $W - 15 = W_0$, serves as a measure of absolute magnitude.

Fig. 7a shows a plot of M_v as a function of log W_0 for 125 low-velocity stars not differentiated by spectral type. They have been divided into three groups on the basis of the intensity of the emission lines. Absolute magnitudes have been inferred from MK luminosity classes through the calibration by Keenan & Morgan. The dashed lines represent the spread corresponding to an error of $\pm 10\%$ in W_0.

Fig. 7b is identical with Fig. 7a except that spectral type is indicated rather than line strength.

Fig. 7c illustrates the relation between M_v (determined from trigonometric parallax) and log W_0. The line drawn is identical with those in Figs. 7a and 7b. The stars included in Fig. 7c are those for which the measured trigonometric parallaxes are at least four times the size of their probable errors.

The evidence strongly supports the hypothesis that the width of the Ca II emission is a function of only the absolute magnitude of the star. Temperature (spectral type) or strength of emission line do not appear to be involved in the relation, which is found to hold over a range of at least 15 magnitudes. Internal consistency of the measurements indicates that one good spectrogram should fix the absolute magnitude of any late-type star with suitable lines to within ± 0.5 magnitude. No satisfactory theory of the observed relation between log W and log L yet exists.

Wilson's method of deriving absolute magnitudes appears, at present, to provide, particularly among the very luminous stars, the highest resolution of any method known. It should make possible a much better picture of the galactic distribution of late-type stars than we have at present. It should find application in a great many fields of galactic research; it may be an aid in the calibration of the absolute magnitudes of other types of stars through its application in galactic clusters.

59

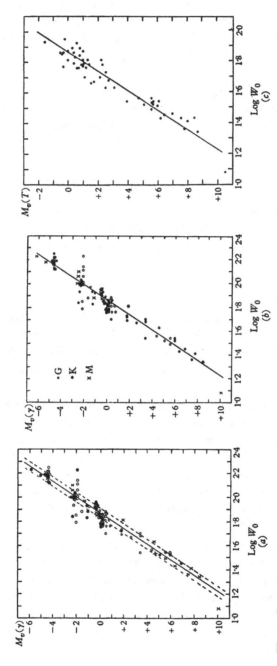

Fig. 7a. Logarithm of the corrected Ca II emission-line widths (W_0) plotted against Yerkes spectroscopic absolute magnitude. Stars divided into three intensity groups. Intensity increases in order 1, 2, ..., 5. The dashed lines indicate the spread corresponding to ±10% error in the measurement of the line width.

Fig. 7b. Same as Fig. 7a except that stars are grouped by spectral type G, K or M. Line intensities not indicated. Straight line identical with that shown in 7a.

Fig. 7c. Logarithm of corrected Ca II emission-line widths plotted against absolute magnitudes derived from trigonometric parallaxes. Straight line same as that in Figs. 7a, 7b.

Wilson plans to extend the study by employing smaller dispersion, 38 Å/mm, and to speed up the measurement process by use of photo-electric scanning technique. He proposes to concentrate, at first, on the study of super-giants, giants, and sub-giants.

(7) *Parenago's plan for co-operative observational work in certain areas*

Parenago discussed a plan for extensive co-operative work on five areas of the Milky Way in the manner proposed by Kapteyn in his Plan of Selected Areas [45]. Effectively, the plan proposed by Parenago [46] would constitute a revision of Kapteyn's Special Plan [45]. It would in no way affect the Kapteyn General Plan.

In the light of modern galactic research, regions of the Milky Way larger in area than those marked out by Kapteyn must be investigated. Each of these larger areas should hold some special interest; each should be chosen carefully on the basis of our best present knowledge of galactic structure. As knowledge increases, or as special needs arise, other areas for co-operative work may be added. In this sense Parenago's plan is a basic one, capable of expansion as needed.

The five areas selected by Parenago for co-operative study are listed in Table 5. Exact boundaries are given in Parenago's paper [46].

Table 5. *Areas in Parenago's Plan*

Area no.	Constella- tion	R.A. 1900	Decl. 1900	centre	b centre	Area in square degrees	Expected no. of stars brighter than $m_{pg}=13.0$
I	Aquila	$18^h\ 32^m$ to $19^h\ 44^m$	$-1°$ to $+18°$	12°	0°	240	22,000
II	Cygnus	20 38 to 20 56	$+43$ to $+47$	53	0	12	2,000
III	Taurus	4 00 to 5 04	$+16.5$ to $+31.5$	141	-13	130	4,000
IV	Taurus	5 22 to 6 02	$+23.5$ to $+30.5$	150	0	45	3,600
V	Orion	5 00 to 5 40	-10 to $+5$	178	-17	143	10,000
Totals						570	41,600

For all stars brighter than photographic magnitude 13 in these regions, Parenago stated it was desirable to obtain the following data:

1. Proper motions. First epoch photographs are available at those observatories that took part in the Carte du Ciel program. The observatories at San Fernando, Algiers, Toulouse, Bordeaux, Paris, Oxford, and Helsinki are involved. Valuable absolute proper motions of long-period Cepheids, clusters, and planetary nebulae are expected by Parenago since the transformation from relative to absolute would be made through a large number of stars in the area. Assistance in measurement and reduction of plates in this extensive program was offered by Parenago.

2. Apparent magnitudes in the photographic and photovisual wave-

length ranges. Later, if possible, apparent magnitudes in the ultra-violet and infra-red should be determined. Magnitudes to a fainter limit may be required in smaller areas within the large regions.

3. Spectral types, with estimates of luminosity classes or absolute magnitudes.

4. Polarization measurements.

5. Radial velocities determined by objective prism methods.

6. Studies of known variables, combined with a search for, and subsequent study of, new variables and emission-line stars.

The estimated number of stars to be observed in each region is given in Table 5.

The observational data, when complete, will permit many studies to be made, among them being the following:

1. Determination of the extinction law by various methods: (a) star counts, (b) color excesses, (c) polarization, (d) mean proper motions in different magnitude intervals, (e) three-color methods.

2. Determination of the space density from star counts according to apparent magnitude.

3. Investigation of components of proper motion perpendicular to the galactic plane. This investigation is possible since three of the five regions lie on the galactic equator. In these regions, the z-components of stars can be derived without radial velocities.

4. In the region in Cygnus a study of components perpendicular to the direction of galactic rotation can be made; in the fourth region, in Taurus, a study of components in the direction of rotation is possible. Radial velocities will not be necessary in either case.

5. In the second through fifth regions stellar associations can be investigated.

6. Nearby obscuring clouds occur in the first and third regions.

Much discussion followed Parenago's proposal. There was general agreement that the proposed plan was an interesting one, but reservations were expressed in regard to the possibilities of carrying it through to completion. Vyssotsky especially stressed the danger of serious systematic errors in proper motions derived from Carte du Ciel plates. Blaauw, Oort, and Malmquist mentioned the need for separation of different population components in kinematic studies. Such separation is not basic to the proposal as it was outlined; but to include it would increase the work significantly. At the Groningen Symposium on Co-ordination of Galactic Research it was emphasized, for example, that it is of greater value to have accurate spectrographic data for a small number of stars than just star counts, even if in two colors, for a large number of stars. Smaller areas than those proposed might be advisable. Lindblad and Schalén stressed the large amount of labor involved in the plan and called attention to work now in progress on at least some aspects of such area surveys in other parts of the sky. More than five areas are required if we are to gain a well-rounded picture of the Galaxy; if more numerous areas are investigated, each may have to be smaller than the five suggested.

Parenago agreed that the determination of proper motions would be difficult, but expressed the belief that only Carte du Ciel observatories could determine adequate proper motions since only they can have two epochs separated by sixty years. He stated that the proper motions should be determined with the help of AGK 2 or 3. He agreed fully that the separation of the stars into different population components was most important. Possibly the general study of the five areas should be restricted to stars earlier than A5. More areas may be required, but, in any case, so many observations are required that a co-operative plan of the nature suggested is called for.

(8) *Remarks on spectral surveys*

Kharadze called attention to the importance of the study of A-type stars. At the Abastumani Observatory 55,000 HD A-stars have been mapped; the galactic belt between $b = \pm 10°$ has been intensively investigated. Photometric distances were computed; the effects of interstellar extinction were removed by use of Parenago's model[47] of the reddening medium. About twenty clusterings of A-stars were found. Three of these coincide in direction with O associations; one coincides in direction with a T association. The average size found for an A-star group is $40 \times 40 \times 60$ pc, with the long dimension in the line of sight. This elongation is fictitious; it occurs because of uncertainties in the distances.

A stars and their importance in galactic structure problems were extensively discussed in the first Symposium on Co-ordination of Galactic Research. The need for studies of the space distribution and kinematic studies of these stars remains. Morgan and Thackeray noted that serious regional systematic errors in classification of A stars are present in the HD catalog. Care must be exercised in using the HD spectral types in large surveys.

Morgan mentioned the desirability of extending to fainter magnitudes the M-star survey[48] made by Nassau and his associates at the Warner and Swasey Observatory. M stars are among the main indicators of the disk population of the Galaxy. A catalog listing, possibly, the 2000 brightest members of this group and extending around the entire galactic belt would be exceedingly valuable. There is a possibility that the Stockholm

Observatory, using the equipment at the Boyden Station, will take an active part in a southern survey of these stars.

At the Leander McCormick Observatory spectral types are being determined for stars in the AGK 2 and AGK 3 catalogs. From the observations to date three facts have emerged: (a) the spectral types listed in *Yale Transactions*, vol. 24, are systematically late. Such an error can be the source of false statistics and spurious effects among the fainter stars. (b) Some stars show simultaneously weak hydrogen and H and K lines of Ca II. There is some tendency for these objects to appear in bunches. (These are probably metallic-line stars, though the explanation needs investigation.) (c) Certain other stars also appear to have a clumpy distribution—the F8 stars, for example; objects rare in the HD catalog.

References

[1] I.A.U. Symposium No. 4, *Radio Astronomy*, edited by H. C. van de Hulst, Cambridge University Press, 1957; see also *Bull. Astr. Inst. Netherl.* **13**, no. 457, 1957.

[2] See Ollongren, A. and van de Hulst, H. C. *Bull. Astr. Inst. Netherl.* **13**, 196, 1957; Westerhout, G. *Bull. Astr. Inst. Netherl.* **13**, 201, 1957; Schmidt, M. *Bull. Astr. Inst. Netherl.* **13**, 247, 1957.

[3] Blaauw, A. *Bull. Astr. Inst. Netherl.* **11**, 459, 1952.

[4] See van de Hulst, H. C. I.A.U. Symposium No. 4, p. 5, 1957.

[5] Communicated by F. J. Kerr.

[6] Kwee, K., Muller, C. A. and Westerhout, G. *Bull. Astr. Inst. Netherl.* **12**, 211, 1954.

[7] Schmidt, M. *Bull. Astr. Inst. Netherl.* **13**, 15, 1956.

[8] See the diagram by Schmidt, M. *Bull. Astr. Inst. Netherl.* **13**, 247, 1957.

[9] Kerr, F. J., Hindman, J. V. and Carpenter, Marther Stahr. *Nature*, in press.

[10] Lindblad, B. *Stockholms Obs. Ann.* **19**, no. 9, 1957.

[11] Weaver, H. F. *Smithsonian Cont. to Astrophysics*, **1**, 151, 1956.

[12] Kerr, F. J., Hindman, J. V. and Carpenter, Martha Stahr. *Nature*, in press.

[13] Kerr, F. J. *Astron. J.* **62**, 93, 1957; also found by Burke, B. F. *Astron. J.* **62**, 90, 1957, and by Westerhout, G. *Bull. Astr. Inst. Netherl.* **13**, 201, 1957 (Fig. 7).

[14] Schwarschild, M. *A.J.* **59**, 273. 1954.

[15] Westerhout, G. *Bull. Astr. Inst. Netherl.* **13**, 201, 1957.

[16] van Tulder, J. J. M. *Bull. Astr. Inst. Netherl.* **9**, 315, 1942.

[17] See the report on p. 39 in this volume; also an account given in I.A.U. Symposium No. 1. *Co-ordination of Galactic Research*, edited by A. Blaauw, Cambridge University Press, 1955; Nassau, J. J. and Blanco, V. M. *Astroph. J.* **120**, 118, 465, 1954; Nassau, J. J., Blanco, V. M. and Cameron, D. M. *Astroph. J.* **124**, 522, 1956.

[18] For more extended surveys of specific topics see I.A.U. Symposium No. 4.

[19] Piddington, J. H. and Trent, G. H. *Aust. J. of Phys.* **9**, 481, 1956.

[20] Westerhout, G. I.A.U. Symposium No. 4, p. 22; discussion of remarks by R. Hanbury Brown, I.A.U. Symposium No. 4, p. 216, 1957.

[21] Unpublished material supplied by T. E. Houck.

[22] Morgan, W. W., Strömgren, B. and Johnson, H. M. *Astroph. J.* **121**, 611, 1955; Abt, H. A., Morgan, W. W. and Strömgren, B. *Astroph. J.* **126**, 322, 1957.

[23] Courtès, G. *Mem. Soc. R. Sc. Liège*, 4ᵉ serie, **15**, 453, 1954.

[24] Becker, W. *Astroph. J.* **107**, 278, 1948; Johnson, H. L. and Morgan, W. W. *Astroph. J.* **117**, 313, 1953.

[25] Morgan, W. W., Meinel, A. B. and Johnson, H. M. *Astroph. J.* **120**, 506, 1954; Schulte, D. H., *Astroph. J.* **123**, 250, 1956.
[26] *Cape Mimeograms*, no. 2, no. 3, 1953.
[27] Shapley, H. *Proc. Nat. Acad. Sci.* **26**, 541, 1940.
[28] Blaauw, A. and Morgan, H. R. *Bull. Astr. Inst. Netherl.* **12**, 95, 1954; Weaver, H. F. *Astron. J.* **59**, 375, 1957.
[29] Code, A. D. *Astroph. J.* **106**, 309, 1947.
[30] Feast, W. M. *Mon. Not. R. Astr. Soc.* **116**, 583, 1956.
[31] A large number of values were considered: Stoy, R. H. *Vistas in Astronomy*, vol. 2, 1956, Pergamon Press, London, p. 1099; Greenstein, J. L. and Henyey, L. G. *Astroph. J.* **93**, 327, 1941; Stebbins, J. *Observatory*, **70**, 206, 1950; Morgan, W. W., Harris, D. L. and Johnson, H. L. *Astroph. J.* **118**, 92, 1953; Hiltner, W. A. and Johnson, H. L. *Astroph. J.* **124**, 375, 1956; Blanco, V. M. *Astroph. J.* **123**, 64, 1956.
[32] Joy, A. H. *Astroph. J.* **86**, 363, 1937; Stibbs, D. W. N. *Mon. Not. R. Astr. Soc.* **115**, 363, 1955.
[33] Morgan, W. W., Whitford, A. E. and Code, A. D. *Astroph. J.* **118**, 318, 1953.
[34] Eggen, O. J., Gascoigne, S. C. B. and Burr, E. J., to be published in *Mon. Not. R. Astr. Soc.*
[35] Stibbs, D. W. N. *Mon. Not. R. Astr. Soc.* **115**, 323, 1955.
[36] Unpublished observations by T. Schmidt of Göttingen.
[37] Kraft, R. P. *Astroph. J.* **126**, 225, 1957.
[38] van den Bergh, S. *Astroph. J.* **126**, 323, 1957.
[39] Eggen, O. J. *Astroph. J.* **113**, 367, 1951.
[40] Irwin, J. B., unpublished.
[41] Joy, A. H. *Astroph. J.* **86**, 363, 1937.
[42] Stibbs, D. W. N. *Mon. Not. R. Astr. Soc.* **115**, 363, 1955.
[43] Wilson, O. C. and Bappu, M. K. Vainu. *Astroph. J.* **125**, 661, 1957.
[44] Wilson, O. C. *Proc. N.S.F. Conference on Stellar Atmospheres*, edited by M. H. Wrubel, 1954.
[45] Kapteyn, J. C. *Plan of Selected Areas* (1906). Reprinted 1923, Hoitsema Brothers, Groningen.
[46] Parenago, P. P. *Astr. J. U.S.S.R.* **33**, 749, 1956.
[47] Parenago, P. P. *Astr. J. U.S.S.R.* **22**, 129, 1945.
[48] Nassau, J. J. and Blanco, V. M. *Astroph. J.* **120**, 118, 464, 1954; Nassau, J. J., Blanco, V. M. and Cameron, D. M. *Astroph. J.* **124**, 522, 1956.

(G) LOCAL STRUCTURE AND STELLAR MOTIONS

The region within about a kiloparsec of the sun contains no globular clusters, and few associations, planetary nebulae, novae, or Cepheids; the interpretation of 21-cm observations is complicated by peculiar motions which are comparable with galactic rotation effects. Nevertheless, this is an extremely important region for the study of galactic structure. Not only are all distance calibrations ultimately based on objects near the sun, but only in this region do we have the resolution necessary to examine the detail of galactic structure and only here can we study the numerous fainter members of the various stellar populations. As elsewhere, the problems are two-fold, the efficient selection of those objects whose study is the most informative and the choice of the methods of obtaining the greatest amount of information about those objects.

(1) *Spectral surveys*

The Kapteyn Selected Areas provide a simple selection of objects for investigation and are small enough to permit a study of all of the stars down to a comparatively faint limiting magnitude. Elvius reported on his investigation [1] of a number of selected areas. He also compared his results with those of other observers. Objective prism spectra were obtained for stars down to about 13·5 photographic magnitude and were classified on the Stockholm system. There is a good correlation between these types and those in the Bergedorfer Spektral Durchmusterung for the brighter stars, but for those fainter than 12th magnitude, the Stockholm classifications appear more accurate. Vyssotsky mentioned that at Virginia they had found that large errors in the Bergedorf Spektral Durchmusterung were usually the result of mis-identifications. Both Virginia and Stockholm classify the K stars a little later than Bergedorf, as a result of using two plates, one for the H and K region and one for the region near λ 4400. Among the F stars, various classification systems differ according to whether they use the intensity of the hydrogen lines or the ratio of Hγ to the G band. Weak-line stars will be classified later from the former criterion. Elvius also found that relations between various spectral systems vary with the luminosity of the stars compared. The Stockholm system is a consistent one, however, and shows no change with either observer or time. It can be converted to the MK system with an accuracy of three or four tenths of a spectral type. A study of population classes will be made later.

Elvius' investigations were based on the Stockholm magnitudes, which are very close to the international system of pv and pg magnitudes. For the mean of eight or nine plates, the internal accuracies are 0·03, 0·04, and 0·08 magnitudes for the photovisual and photographic magnitudes and the colors, respectively. The intrinsic colors of stars of each spectral type were determined on this system and agree excellently with those determined by Ramberg. Malmquist reported that at the north galactic pole, the spectral classification determines the color of a star within one or two hundredths of a magnitude (determined photo-electrically). His colors for the giants agree well with those of Elvius, but the main sequence stars are between one and two tenths of a magnitude bluer than the Stockholm intrinsic colors. In future programs, ultra-violet magnitudes should also be measured.

66

For each spectral type in each region, a plot was made of color excess against apparent distance modulus. These plots showed definite deviations from linear relations. While often not significant individually, deviations were usually at the same distance in plots for various groups of stars, indicating that they resulted from true variations in absorption and star density. Moreover, neighbouring regions often showed similar variations at the same height above the galactic plane, even though the line-of-sight distances to the variations were decidedly different. There were rather definite indications of higher star densities over the Orion arm and lower densities in the inter-arm regions. In some regions, extended absorption clouds could be traced to large distances above the galactic plane. The poor location of the selected areas in the direction of the Perseus arm (in regions of high interstellar absorption) made the study of that arm difficult. A ratio of four was used to convert selective absorption to total absorption.

(2) *Search for sub-giants and M dwarfs*

Except in very small regions of the sky, it is impossible to study all stars in detail; some method must be used to segregate those of special interest. Eggen outlined the method by which sub-giants within 20 pc are being selected for the Greenwich parallax programme. All K 0 stars north of the equator in the Henry Draper Catalogue between magnitudes 5·5 and 8·0 are examined spectroscopically. Stars brighter than this limit have already been classified on slit spectra; fainter sub-giants are too distant. Later, the program will be extended to G stars and late F stars, but these are more difficult to classify accurately. Sandage emphasized that this is a very important program since by studying the faintest sub-giants in the solar neighborhood we will be able to determine theoretically the age of the oldest stars in this region. Vyssotsky described the failure of the McCormick program for determining the parallaxes of sub-giants selected by proper motions. A few of the stars were dwarfs, but most had parallaxes comparable with their errors and it was impossible to determine their luminosities. The Greenwich program will avoid this difficulty by selecting spectroscopically only stars whose parallaxes are greater than $0''05$. Moreover, Eggen emphasized that parallaxes should always be determined at more than one observatory. Greenwich will remeasure the parallaxes of all stars known to be nearer than 20 pc for which only one parallax determination exists, as well as those stars whose parallaxes measured at two observatories differ by a factor of two or more. A list of the stars on their parallax program will be distributed to anyone interested.

Fricke proposed that very low dispersion spectra, taken, for example, with a small Schmidt telescope and an objective prism, be used to segregate the M stars. In regions near the galactic poles, many of these would be nearby dwarfs, although Nassau pointed out that the 300 square degrees at the pole on the Cleveland survey contain only thirty-five M stars of all luminosities. By using infra-red plates proper motions could be measured for these stars to a limit of 20th photographic magnitude. The scale of the Hamburg Schmidt would be sufficient to provide usable proper motions within a few years. Lindblad pointed out that at 2000 Å/mm, it is possible to distinguish between M giants and M dwarfs, certainly among stars as faint as 15th magnitude and possibly among stars as faint as 17th magnitude.

(3) *High-velocity stars; separation of population groups*

Miss Roman reported on some of the results of her study of high-velocity stars. Most of the six hundred stars which she studied are obviously members of the disk population. Except for two small groups, the average velocity of the stars perpendicular to the galactic plane is little higher than the average Z velocity for all of the weak-line stars near the sun. The two groups excepted are particularly interesting. The F-type sub-dwarfs have very high Z velocities and probably would be found very high above the galactic plane if they were not too faint to be discovered at large distances. The second group, the fifteen high-velocity A stars, is a more surprising one. The velocity of any individual A star is not well determined because it depends on the proper motion and parallax, but many of these stars have very large radial velocities and there is no question that most of them are high-velocity objects. The Z velocities for these stars are almost as high as those for the sub-dwarfs and the A stars are much more concentrated to high-galactic latitudes than are the later F stars.

Further evidence that most of the high-velocity stars and the globular-cluster stars belong to different populations is provided by the spectra and by the color-magnitude diagrams of the two groups. Sandage, on the basis of the trigonometric parallaxes, and Miss Roman, on the basis of spectroscopic parallaxes, both had diagrams which showed that the color-magnitude diagram of the high-velocity stars is similar to that of the old open cluster, M 67, or to that of the field stars near the sun and decidedly different from those of globular-cluster stars. A check on the mean parallaxes for small ranges of spectral type and luminosity showed that the use of spectroscopic parallaxes for the high-velocity stars is legitimate and the similarity of the results from trigonometric and spectroscopic parallaxes substantiates this. A comparison of the spectra of bright red giants in three globular clusters with spectra of the most extreme high-velocity stars indicates that the latter are about a luminosity class fainter. Also other differences indicate that the high-velocity stars are more closely related spectroscopically to the stars near the sun than they are to the globular cluster giants. Parenago's result that the high velocity stars on his diagram obey the color-luminosity relation for the globular clusters does not contradict these results as he was considering an extreme group with large Z-velocity components.

The position of the high-velocity stars on a two-color diagram is also interesting. Johnson and Morgan[2] have shown that on a plot of U-B as ordinate against B-V as abscissa, the normal stars fall on a narrow sequence. The high-velocity stars fill a band about 0·2 magnitude wide extending upwards from the normal sequence to that defined by the stars in M 3. If one defines the height of the star above the normal sequence as its ultra-violet excess, U_{ex}, there is an excellent correlation between this ultra-violet excess and the space velocity of the star in the range of spectral types from F o to early K. This correlation is at least as good for each of the components of the total velocity. Both Parenago and Miss Roman emphasized that this indicates a good correlation between U_{ex} and the perigalactic distance. The ultra-violet excess provides a powerful method for selecting members of the disk and halo among faint stars, as well as for segregating dynamically similar groups of stars without biasing the results by using the motion for selection. Either a fourth color or a rough spectral type should be used to correct the measured colors for interstellar reddening, but since the reddening line and the normal star sequence are not far from parallel in this region of the two color diagrams, the ultra-violet excess is not unduly sensitive to reddening.

The long discussion which followed this report was primarily restricted to three problems: the physical explanation of the ultra-violet excess, the use of the ultra-violet excess for selecting disk and halo members, and the problem of locating halo stars. The latter are comparatively rare; if the statistics for the halo in general are similar to those for the globular clusters, the K-type halo giants have about the same space density as the R R Lyrae stars. Miss Roman suspects that the ultra-violet excess results from the weakening of the metallic lines and the cyanogen since both features are more prominent in the ultra-violet than they are in the blue region of the spectrum. Chalonge added that in the intermediate types a decrease in the Balmer discontinuity also influences U_{ex} in the same direction and should not be overlooked. At Mount Wilson, high dispersion spectrograms of six normal standard stars and two abnormal high-velocity stars are being studied to determine if the ultra-violet excess can be explained entirely by the weakening of the spectral lines. The equivalent widths of all of the lines between 3200 Å and 6000 Å are being measured.

Miss Roman suggested that one of the Schmidt telescopes be used to select stars in the range A 0 to K 0, or in a narrower range of special interest. Later than K 0, luminosity effects become hard to handle among the giants, and the dwarfs are too close to be interesting. For each of these stars, three-color photo-electric photometry should be obtained. Those stars with U_{ex} (corrected for absorption) greater than 0·08 magnitude will be a relatively small sample of stars from either the halo or the disk. Then, for these, radial velocities and proper motions will be very valuable. In the selected areas we already have proper motions for stars as faint as 14·5 photographic magnitude and radial velocities are being determined by Fehrenbach for stars at least as faint as 12th magnitude. Thus these would be interesting regions in which to begin such a programme. A start is furnished by Miss Roman's work. She has spectra of about 600 stars between 8th and 12th magnitude in the selected areas listed in Table 6. Most of the 'fundamental stars' are BD stars and hence brighter than 10th magnitude. Blaauw pointed out that the accuracy of the proper motions could be increased significantly by a repeat of the Radcliffe photographs. Another related problem is a check on the luminosity calibration of the high-velocity giants. The evidence at present is that the normal luminosity calibration can be used, but this should be checked by trigonometric parallaxes for the most extreme weak-line stars. While individual parallaxes will probably be too small to be meaningful, the mean of several determinations for several stars will be valuable. Miss Roman will be glad to suggest a list of stars for a parallax program.

Oort pointed out that of the stars on Miss Roman's program, only a few will be halo objects; most will be main sequence G stars. To avoid this difficulty, it will be necessary to search over a much larger area and to segregate only the red giants for further study. At 13th magnitude, these giants will be 1000–2000 pc from the galactic plane and many will belong to the halo population, or at least to the high velocity population. Therefore, Oort recommends a systematic survey for K giants in a fairly large region around both the north and the south galactic poles to as faint magnitudes as possible. It is important to have a comparison of these with nearby giants on exactly the same system. There are probably only about 100 halo giants in each of the polar caps so it will be difficult to find a significant number.

Several studies of giants in the polar regions are in progress. G. Münch has searched plates taken with the Tonantzintla Schmidt for giants fainter than 11th magnitude in high galactic latitudes. Lindblad stated that it is easy to separate giants and dwarfs among stars as faint as magnitude 13·5 with the Stockholm equipment. The weakness of the cyanogen causes no difficulty as the difference between populations is small compared to the difference between giants and dwarfs. Stockholm observers may also be able to measure radial velocities on their plates. Malmquist is investigating about 64 square degrees around the north galactic pole. He has obtained spectra and color indices for stars down to magnitude 13·5 and photo-electric magnitudes and colors for stars brighter than 10th magnitude. There are about 150 K giants in this material and it is obvious that there is a maximum in the apparent distribution at about 10th magnitude. Sandage recommended photo-electric photometry for the fainter stars to segregate those with large ultra-violet excesses. The decrease in the number of K giants for the fainter magnitudes is encouraging since it means that in the range 14th to 16th magnitude, most of the K giants will probably be members of the halo. Edmondson, of Indiana University, has spectra of 700 K stars brighter than 12th magnitude which he has measured for radial velocity. In high latitudes, not more than 7 % of these stars are dwarfs and at low latitudes, the percentage drops to 3 or 4 %. The stars at high latitudes number only about 80–100 but these have proved useful for the determination of the average velocity of objects at about 1000 pc from the galactic plane. They seem to have larger velocity dispersions than those at lower latitudes although the result is not very definite. However, the average Z velocity is only about 25 km/sec which, while higher than the average for stars near the plane, is not nearly that typical of the halo. The spectra are not suitable for accurate classification but it would be valuable to obtain photo-electric photometry and possibly accurate spectral types for these stars, particularly for those in high galactic latitudes.

Table 6. *Selected areas in which stars have been observed spectroscopically by Miss Roman*

(a) Areas in which the 'Fundamental Stars' have been observed. These stars were observed by meridian circle observers as standards for the determination of positions for the remaining area stars. They are listed in *Leiden Annals*, no. 15.

Areas 5, 13–15, 29–35, 53–60, 64, 74, 78–83, 87*, 92, 93, 98*, 102–106, 110, 115–119, 128, 138–143, 162 and 163.

(b) Areas in which the observations cover in addition to the 'Fundamental Stars' all of the stars to 12th photographic magnitude in the *Radcliffe Catalogue* of proper motions in Selected Areas or in *Harvard Annals*, no. 102.

Areas 9, 13, 29, 31, 32, 34, 55–58, 64, 74, 80–83, 87, 93, 98, 104, 110, 115, 116, 119, 129* and 138.

* Observations incomplete.

Morgan recommended a spectroscopic survey, star by star, with accuracy equal to that of the MK standards for stars, say from the 10th to the 12th magnitude. This would permit a separation of both spectral and population types and the ultra-violet excess could then be studied as an independent parameter. One should use widened spectra with a dispersion near 100 to 200 Å/mm. These should be compared with exactly similar spectra of standard stars taken with the same equipment. The selection effects should also be carefully studied. Morgan thinks that such a program is feasible but stressed that the main problem is one of high systematic accuracy so that stars in the polar cap can be compared with those in the neighborhood of the sun for which we have much more detailed information. If a slit spectrograph similar to that used for the Yerkes standard stars is employed, the systematic accuracy at K 0 III should be of the order of 0·1 of a class. Under the best conditions, with the best plates, the accidental accuracy should be of the same order. The problem of the intrinsic dispersion in luminosity of the K 0 III stars should also be studied. Halo giants might be extremely rare in a sample of giants between 10th and 12th magnitude, but those which are included would be easily recognized spectroscopically. After the possibilities of this method have been exhausted, one can study the ultra-violet excess and use it, perhaps, to extend the search to fainter stars. Morgan recommended that objective-prism spectra be used for segregating the original group of giants near class K 0.

There was some discussion of the limiting magnitude at which the search for halo stars should be conducted. Sandage thought that the halo giants might outnumber the disk giants at high latitudes in the range between 13th and 16th magnitude, and that therefore the search would be most advantageously conducted among stars of this brightness. Morgan thought that 13th and 14th magnitude stars at a distance of 3000 pc and 5000 pc, respectively, were better objects to start with than stars at 16th magnitude. He also stressed the advantage of studying the brighter stars carefully before going to the faintest ones. It would be possible to pick out 16th magnitude stars with color indices between 0·8 and 1·0 with a micro-Schmidt but one could not separate dwarfs and giants in this way. With a Schmidt, which transmits the ultra-violet, it is possible to separate giants and dwarfs with a dispersion of 500 Å/mm (limiting magnitude near 14th), but otherwise a dispersion of at least 300 Å/mm is needed.

Vyssotsky reported on the McCormick study of nearby stars[3]. These included stars between A 0 and A 3 and K stars brighter than 5^{m}5 as well as the dwarf M stars included in the McCormick survey without regard for parallaxes and proper motions. For the latter either trigonometric or spectroscopic parallaxes are available. These stars were grouped in several ways. First, the maximum Z-velocity observed for any of the A stars was used as a criterion to divide the F- and M-type stars into two groups each. Those with small Z move in orbits near the galactic plane; the orbits of the others are inclined 15° to 20° to this plane. On plots of the velocities of the stars in each of these groups, it was found that for the stars with low orbital inclinations, the vertex of the velocity ellipsoid deviates from the direction to the galactic center by 15° or 20°. Conversely, the velocity vectors for the stars with highly inclined orbits show a much larger dispersion but no deviation of the vertex, which Vyssotsky attributes to the presence of the spiral arms. He also divided the stars into groups according to spectroscopic criteria. For the G stars he used the McCormick data on the relative strength of the Hγ line and the G band. Among the K stars, he used only Miss Roman's material which separated the strong- and weak-line stars. The basis of separation in the M's was whether or not the stars show emission lines. The latter is not as clear a separation as it might be because the spectrograms which were obtained at Mount Wilson tend to be somewhat underexposed. The plots of the velocity vectors for these stars show the same characteristics as the plots for the earlier separation. The A stars, the strong-line stars, and the M dwarfs with emission lines have velocity vectors which show a small dispersion but for which the major axis of the velocity ellipsoid deviates strongly from the direction toward the galactic center. Conversely, the weak-line G and K stars and the M dwarfs without emission lines show a larger dispersion but no deviation of the vertex. It would be useful to do an analysis of this sort of the AG stars in the 20° to 25° zone for which radial velocities have been published by the David Dunlap Observatory [4]. There are also recently improved proper motions for these stars. A separation into strong- and weak-line stars would be necessary in addition.

(4) *Stars with hyperbolic velocities*

According to Perek [5], if the circular velocity near the sun is 216 km/sec and the escape velocity exceeds this by 65 km/sec, then about sixteen stars near the sun have hyperbolic velocities. There are about six sub-dwarfs, three or four RR Lyrae-type stars and the youngest star is AE Aurigae. Also included are the B I IV star BD + 28° 4177, van Maanen's star, a white dwarf, AG Aurigae, an RV Tau star, and a globular cluster, NGC 5694. The proper motion of the latter is extremely uncertain but the hyperbolic velocity depends only on the radial component. Three assumptions enter the selection of stars with hyperbolic velocities: the value of the circular velocity, the value of the velocity of escape, and the assumption that the spectral lines really indicate the motion of the center of mass of the star or star system. Oort estimates that the uncertainty in the circular velocity is at least 10 % and possibly larger. The velocity of escape is still more uncertain because so little is known about the density in the outer shells of the galactic system.

Perek suggests that a survey for large radial velocities is the most effective way to find stars with hyperbolic orbits. For instance, among thirty-eight stars with radial velocities greater than 250 km/sec there are many hyperbolic velocities but a direct elliptical motion can be established for only one. Fehrenbach already has a program for finding large radial velocities on his objective prism spectrograms. Many stars may well have hyperbolic orbits but the observational data necessary to decide this are lacking. For example, about seven stars need a new determination of the radial velocity, eleven need a new determination of the parallax, and for about forty, new proper motions are desirable. Blaauw emphasized the importance of sub-dividing stars with hyperbolic orbits into those which are apparently very young and have been formed with very high velocities within our Galaxy and other stars which, with reasonable assumptions about their age, must have been formed outside the Galaxy. It is reasonable to expect such interlopers. If we consider a star like AE Aur, we find a velocity of between 120 and 130 km/sec with respect to its origin in the Orion region. This star happens to have a velocity about 30 km/sec larger than the velocity of escape as presently estimated, but if it were formed in a stellar system where the total mass were much smaller, there would be no question but that it would become an intergalactic object. The number of these stars which can be expected depends on the rate of formation, which might be quite large in the Magellanic Clouds, as well as on the gravitational field within the system. Hence, the Magellanic Clouds may well be surrounded by a halo of young high velocity stars.

(5) *Radial velocity programs*

Fehrenbach reported on the objective-prism radial velocity program carried on at the observatories at Haute-Provence and Marseille. So far, nearly 1300 plates have been taken, the first 400 of which were 9 × 12 cm across. This size is no longer used as it has been found that the usable field is actually much larger. It was first replaced successively by 13 × 18 cm plates (about 700 of which were taken) and, during the past year, by 16 × 16 cm plates corresponding to a 4° × 4° field. With the small objective prism of 15 cm diameter, magnitude 10 can be reached with an exposure of twice 2 hours. In a field of $4° \times 4°$ near the Milky Way, between 80 and 120 stars are well measurable.

So far, fifteen fields have been measured in galactic latitudes near zero and at longitudes, separated by 15°, between 345° and 180°. Four fields containing 240 stars have been published [6], and nine fields containing 800 stars are ready for publication. In addition there is a field in Coma Berenices with about fifty stars. In each field three plates are taken with twice 2 hours exposure covering the range 7^m8 to 10^m, and 3 plates with twice 40 minutes exposure covering the magnitudes down to 8^m5. Thus there is an overlap between 7^m8 and 8^m5 on both sets of plates to insure a homogeneous system.

The scarcity of well determined slit radial velocities in these fields is a serious problem which has been brought to the attention of the I.A.U. sub-commission on Standard Velocity Stars. In the B and A stars and the F stars earlier than F 6, usually only $H\gamma$ and $H\delta$ are measured, but in the later stars, types F 6 to M, about eight lines, spread over the spectral region, are used. The latter were chosen after a preliminary study similar to that which is carried out for the choice of lines on slit spectra. The mean error per star for radial velocities based on about six plates is ± 4.8 km/sec for types B, A, F and ± 3.1 km/sec for types G, K, M. Some of the errors seem to be due to displacements of the gelatine on the plates. For this reason the Mount Stromlo announcement of a new development technique was a welcome one [7].

In each of Fehrenbach's fields, spectral classes on the MK system are being determined, partially with the collaboration of Kourganoff at Lille. A large number of the stars which seem particularly interesting are being measured photo-electrically on the U-B-V system at Toulouse.

In addition to these programs, twenty plates covering the P Cygni association have already been classified and the measuring is well advanced. Two regions in Cygnus are being studied, one with P Cygni as its center and the other near 28 Cygni. Further, ten fields have been selected in regions of the Milky Way with low absorption. Selected were those rich in O- and B-type stars on the basis of the classifications by Nassau and Morgan. So far thirty plates have been taken and the measurements are well under way. Finally, plates have been taken in nine additional galactic fields. For the future, a general study of all Selected Areas is planned. About thirty plates have already been taken.

High-velocity stars are searched for, by a special rapid procedure, on each field observed so far.

A new objective prism, with a diameter of 40 cm and a dispersion a little smaller than that which has been used, has been constructed. If used with a Grubb and Parsons triplet objective the limiting magnitude will probably be about 13. This combination will cover a $2° \times 2°$ field and will be sufficient to pick out stars in the Magellanic Clouds; these stars will have a Doppler displacement of the order of 40μ which is easily visible provided the spectra are well exposed.

There is no dependence of the accuracy or of the zero-point of the velocity measurements on the length of the exposure.

Woolley mentioned that Greenwich is obtaining a radial-velocity objective prism for use with the 10-inch Astrograph at the Royal Observatory. He is primarily interested in studying the nearby stars.

(6) *Photometric spectral classifications of high accuracy*

Strömgren has developed a photo-electric method of spectral classification. He uses interference filters to isolate wavelength regions of the order of 35 to 40 Angstroms in width near important spectral features such as the hydrogen lines, particulary Hβ, the Balmer discontinuity, the break in the spectrum at the G band, and the cyanogen absorption [8]. Comparison regions are also measured near each of these wavelengths to correct for both interstellar and atmospheric extinction. These measures define the main sequence well but some scatter remains after allowance for photometric errors; this is undoubtedly due to a third dimension, or population variation, in the spectra.

An instrument is now in use with which the narrow band intensity in Hβ and the intensity in the comparison band are measured simultaneously, which completely eliminates variations in sky transparency. Other line strengths are also measured with different filter pairs. The limiting magnitude with this method, the 82-inch reflector, and an integration time of about 1 minute is near 12th magnitude; with the 200-inch and a longer integration time it should be possible to reach stars of 17th magnitude. At present, the classification of 2000 B and A stars in clusters and associations is in progress as well as the observation of B 8 to F 1 stars within 1000 pc of the sun. The purpose of the latter program is a determination of the dust distribution near the sun; this spectrophotometric method can be used to compute and to eliminate color excesses as well as to determine spectral types and luminosities of the stars. A similar program is planned for determining the dust and star distributions to greater distances in particular Milky Way areas.

Chalonge has obtained extremely high accuracy in spectral classification using a more conventional spectrophotometry of photographic spectra[9]. He measures three criteria in each spectrum. They are: the spectrophotometric gradient in the blue region of the spectrum, the wavelength, λ_1, at which the apparent continuum is half-way between the extensions of the blue and ultra-violet continua, and the magnitude of the Balmer discontinuity. On a three-dimensional diagram defined by these parameters, the normal main sequence population I stars fall on a well-defined surface. From the position of a star on this surface, the spectral type and luminosity can be predicted with more accuracy than the MK classification and on the same system.

It is interesting to compare the F stars in the Hyades and the Coma clusters. The Coma main sequence stars seem to have lower luminosities than the Hyades stars. This also explains the color excess observed for the Coma stars since, having lower luminosities, they also have smaller Balmer jumps and hence are brighter in the ultra-violet than the somewhat more luminous stars in the Hyades. This comparison stresses the importance of the Balmer jump in determining the ultra-violet excess of stars in addition to the importance of the metallic lines. Sandage agreed as to the importance of the Balmer jump difference between the two clusters. He believes that the Coma cluster is a younger cluster and is in a slightly different stage of evolution than the Hyades. For this reason the F stars in Coma have slightly higher surface gravities than those in the Hyades.

Compared to the normal stars, the F-type sub-dwarfs of the same gradient have decidedly smaller Balmer discontinuities and slightly smaller λ_1's. Conversely, the metallic-line stars have larger Balmer jumps and larger λ_1's. There is a continuous distribution of stars from the most extreme sub-dwarfs through the normal F-type stars to the most pronounced metallic-line stars. Moreover, the spectrophotometric indication of the degree of peculiarity in these stars is in agreement with estimates from other methods. For example, the high-velocity stars which have not been classified as sub-dwarfs seem to be intermediate between the extreme sub-dwarfs and the normal main-sequence stars. The very small Balmer discontinuities observed in the sub-dwarfs explain the strong ultra-violet excesses of these stars almost completely. There is also a small difference between the shapes of the visible continuum in the sub-dwarfs and in the main sequence stars. The normal stars show a small discontinuity which the sub-dwarfs do not have.

Although the surface, in Chalonge's representation, is well defined for the normal stars it is not infinitely narrow. Probably the spread near the surface corresponds to an intrinsic spread in the character of the stars corresponding, for example, to Miss Roman's division of stars into strong-

77

and weak-line groups. This has been partially tested by observing some of Miss Roman's stars, and the preliminary results agree with this interpretation. The weak-line stars fall on the side of the surface with the sub-dwarfs and population II stars while the strong-line stars, for the most part, fall on the opposite side. It would be interesting and important to study the B stars in the same way but this is more difficult because of the effects of interstellar absorption.

Although the method described is very time consuming it does permit a very accurate classification and it may be a useful way to give some indication of the physical and chemical properties of various stars. Chalonge is willing to study any star brighter than 10th magnitude provided it is sufficiently early in type to have a Balmer discontinuity. The success of this method stresses particularly the importance of using the Balmer discontinuity and the ultra-violet region of the spectrum for the study of classification problems. Madam Hack in Italy is experimenting with the use of photo-electric techniques for this type of classification. She has substituted the intensity of Hβ for the measurement of λ_1 so that her system is also somewhat similar to that of Strömgren. Strömgren's and Chalonge's methods are not too different and Strömgren's measurement of the total intensity of groups of lines will give a third parameter which will be even closer to Chalonge's.

(7) The luminosity function

Recent investigations by Sandage and Salpeter, reported on p. 18 of this volume, have revived interest in the luminosity function. Published luminosity functions such as those by van Rhijn and Luyten include not only main sequence stars but also giants and a few super-giant stars. Different portions of the luminosity function have been determined in different ways. For the fainter stars trigonometric parallaxes have been used; in the region of somewhat brighter stars statistical analyses of proper motions have been valuable; and the brightest end, which is the most uncertain, has been determined primarily from the statistics of stars in the Magellanic Clouds. Blaauw emphasized that the present problem is to redetermine accurately the luminosity function for main sequence stars only. In the region near the sun trigonometric parallaxes can be used to give sufficient statistics for stars fainter than about $+6$. Thus an extension of the trigonometric parallax program is needed. For the brightest stars, say brighter than about $+2$, the best procedure might be a careful redetermination of spectroscopic absolute magnitudes. Such a program should include the stars brighter than about 7·0 and hence an extension of the present classification from 5·5 to 7 would be extremely valuable. In the intermediate region of absolute magnitudes, between +2 and +4, the best course for future study is less obvious. Perhaps Wilson's method of determining absolute magnitudes as a function of the width of the reversal of the H and K line (see p. 59 of this volume) may be the most powerful method for locating stars in the region above the main sequence. This should also be done for all stars brighter than 7·0.

Weaver pointed out that the luminosity function which Blaauw had described referred to different volumes of space in different regions of luminosity. At Berkeley there is an interest in determining the luminosity function for small volumes as, for example, in clusters. These luminosity functions have been determined purely by counting procedures. That is, the number of stars in a cluster has been counted as the excess over the background distribution. In he cluster NGC 7160, a 1b2 cluster on Trumpler's classification, the diagram starts at about absolute magnitude 0 and extends to about +8. There are no detectable faint stars, which may mean that their number is less than the number predicted on the basis of the van Rhijn luminosity function. The 1b6 cluster NGC 7243 extends a little fainter in luminosity but is also comparatively lacking in faint stars.

Heckmann was willing to admit that there may be well established differences in the luminosity functions between a cluster and the region near the sun but he felt that it is important to be very cautious in stating that there are no faint stars in some open clusters. He thinks that there is greater reason to believe that there are some faint stars—many more than can be detected by Weaver's method. For example, when stars are selected by proper motions as in Hyades, Praesepe, Pleiades, Coma and the alpha Persei cluster one can say with certainty that the luminosity function seems to be constant down to an absolute magnitude near +10 and that limit is the limit of the proper motions, not of the stars themselves (see also p. 9 of this volume). However, Baade commented that the luminosity function in star clusters, and in the general field need not be the same. An excellent example is the τ Canis Majoris cluster. Both on Schmidt plates and in more careful investigations, the main sequence of this cluster seems to stop near B8. It is absolutely certain that there is not a large number of faint stars. This is a very young cluster and B8 is very near the point at which the main sequence would be expected to end on evolutionary theories.

Walker has observed three clusters in which the luminosity functions seem to be nearly the same as the 'Initial Luminosity Function' of stars near the sun. These are the Orion cluster, NGC 2264 and NGC 6530 or

M8. McCuskey's investigations[10] of selected Milky Way fields indicated that the luminosity function is quite uniform in general but there are certain regions in which there are too few stars in small intervals of absolute magnitude.

References

[1] Elvius, T. *Stockholms Obs. Ann.* **16**, no. 5, 1951; **19**, no. 3, 1956. The catalogs are published in *Stockholms Obs. Ann.* **16**, no. 4, 1951 and **18**, no. 7, 1955.
[2] Johnson, H. L. and Morgan, W. W. *Astroph. J.* **117**, 313, 1953; *Cont. McDonald Obs.*, no. 216.
[3] See also *Pub. Astr. Soc. Pacif.* **69**, 109, 1957; *Pub. McCormick Obs.* **12**, Part 4.
[4] Heard, J. F. *Pub. David Dunlap Obs.*, Toronto, **2**, no. 4, 1956.
[5] See also *Astr. Nachr.* **283**, 213, 1956; *Veroff. Astr. Inst. Brno*, no. 2; *Bull. Astr. Inst. Czechoslovakia.* **8**, 177, 1957.
[6] Duflot, M. and Fehrenbach, C. *Publ. Obs. Haute Provence* **3**, no. 41, 49, 1956.
[7] Gollnow, H. and Hagemann, G. *A.J.* **61**, 399, 1956.
[8] See also Strömgren, B. *Vistas in Astronomy*, 1336, 1956 (Pergamon Press, London) and *Proc. Third Berkeley Symp. on Mathematical Statistics and Probability*, vol. III, 49, 1956 (University of California Press).
[9] See also Chalonge, D. *Ann. d'Astroph.* **19**, 258, 1956 and *Vistas in Astronomy*, 1328, 1956 (Pergamon Press, London).
[10] McCuskey, S. W. *Astroph. J.* **123**, 458, 1956.

III. THE MAGELLANIC CLOUDS

(A) INTRODUCTION

The Magellanic Clouds offer a unique opportunity to study fully resolved stellar systems at optical and radio wave-lengths. Optically, with the largest southern telescopes available it is possible to reach absolute magnitudes +1 to +2 by direct photography and absolute magnitudes −6 to −4 spectroscopically. The greatest value of the Clouds to galactic research is the opportunity to calibrate the absolute magnitudes of luminous objects within the Galaxy. Uncertainty regarding location in depth within the Clouds adds a random error of about ±0·13 m, i.e. considerably less than the uncertainty of the distance modulus of the Clouds themselves.

In addition, study of the brightest stars may be useful both for calibrating the extra-galactic distance scale, and for picking out, far more easily than in the Galaxy, 'super-super-giants' with absolute magnitudes brighter than −7.

In radio work the chief interest is in the Clouds as galaxies. Data on large-scale properties such as size, luminosity, mass and mass-distribution are required.

An admirable survey of work on the Clouds with bibliography up to about 1954 is contained in the compilation of Buscombe, Gascoigne and de Vaucouleurs[1].

(B) OPTICAL PROGRAMMES

(1) *Photo-electric photometry*

Work on the establishment of sequences in the U-B-V system and on magnitudes of classical Cepheids is being continued at Canberra. The presence of very blue and very red stars in the Clouds indicates the desirability for a wide range of colours in the standard stars chosen for sequences. New sequences, down to magnitudes 18 and 19, in the Small Cloud will become available from the work of H. C. Arp at the Cape and Radcliffe Observatories. Studies of population characteristics, particularly in the Small Cloud, are also under way (H. C. Arp and Mt Stromlo). Three-colour photometry of stars in the general field and in associations is being carried out at the Radcliffe Observatory. NGC 330 is being observed at Mt Stromlo, NGC 2004 and the neighbourhood of S Dor at the Radcliffe Observatory.

(2) *Spectroscopy*

Emission objects discovered on objective prism surveys have been listed by Henize [2] and by Lindsay [3].

Spectra with dispersions of 86 and 49 Å/mm of the brightest stars in both Clouds are being obtained in considerable numbers at the Radcliffe Observatory. Radial velocities of these stars discussed by Feast, Thackeray and Wesselink [4] have confirmed the rotation of the Large Cloud indicated by the radio work and also indicate a moderately large velocity dispersion of the order of 20 km/sec. According to Feast [5] and Feast and Thackeray [6] there are several bright Cepheids and a few red super-super-giants which have been shown to be among the brightest members of the Large Cloud. The H-R diagram of the brightest stars slopes upwards to the right among the B to A types in a manner similar to the apparent evolution tracks of young massive stars in galactic clusters.

Low dispersion, 100–400 Å/mm, spectroscopy has also begun at Mt Stromlo. In particular, emission objects are being studied by G. de Vaucouleurs.

The presence of planetary nebulae in the Small Cloud, with considerable spread in absolute magnitude, has been demonstrated by the spectroscopic work of Lindsay [3] and by photography in selected wave-lengths by Koelbloed [7].

(c) FUTURE NEEDS

Programmes on which further efforts are especially desirable are listed below, in which the first four items refer to the largest telescopes available, while the others would be suitable to smaller telescopes:

(a) Accurate photometry of classical Cepheids and RR Lyrae variables;

(b) Spectral classification and radial velocities of the brightest stars;

(c) Colour-magnitude arrays of associations and globular clusters (including radial velocities where possible);

(d) Searches for long-period, RV Tau and RR Lyrae variables in the general field;

(e) Objective prism surveys for classification and radial velocities, especially in the Small Cloud;

(f) Searches for bright variables (brighter than 13th magnitude);

(g) Identification charts.

It may be added that any optical researches designed to improve the value of the inclination of the main mass of both Clouds to the line of sight would be of special value in the rotational analyses.

The photometry of Cepheid variables in both Clouds is of such fundamental importance that it will probably be undertaken at the Radcliffe Observatory and at the Leiden Southern Station (with the new 36-inch light-collector) in addition to Mt Stromlo. Co-ordination of efforts to avoid an undesirable degree of overlap is necessary.

Many more radial velocities of individual stars are required for a satisfactory comparison of the rotation of the Large Cloud with the radio results on the neutral hydrogen gas.

For a complete census of the brightest stars in both Clouds an objective-prism survey enabling members to be picked out by radial velocities is most urgent. In the discussion, the possibility of Dr Fehrenbach bringing his 40-cm objective prism to the southern hemisphere for this purpose was raised. In the Large Cloud the census of bright *blue* stars is reasonably complete because practically all stars classified in the Henry Draper Extension as O, B or Con are members. The census of bright members of type later than A is probably very incomplete.

The fact that the Henry Draper Extension does not include the Small Cloud means that here our census of bright members is at present restricted to eleven stars in the H.D. Catalogue and to the emission objects listed by Henize[2] and Lindsay[3]. Spectral classification from objective prism spectra extending to a fainter limit than the H.D. Extension is a most urgent problem in the Small Cloud and could also be profitably carried out in the Large Cloud.

The few known red 'super-super-giants' apparently show a tendency to variation in light, probably irregular in cycles lasting months or years. More might be discovered, with quite small telescopes, with exposures suited to the magnitude range 10–13.

Studies of colour-magnitude arrays of associations and globular clusters in the Clouds are of paramount importance to current ideas of stellar evolution. The wealth of such objects—and their bewildering variety may not have been fully appreciated—implies that it should be easy for those engaged in such work to co-ordinate their programmes without overlap. Radial velocities of the globular clusters are also urgently required.

All workers on the Clouds with large telescopes are familiar with the difficulty of identifying individual stars in rich fields, more especially when only X, Y co-ordinates are available, as with the Harvard Cepheid variables. Through the co-operation of Professor Malmquist and the Canberra workers a number of charts of both Clouds, marked with X, Y co-ordinates, obtained with the Uppsala Schmidt at Canberra are likely to become available. Such identification charts will be of immense aid to many workers. It may be mentioned that the extensions to the Clouds shown by the radio work will probably imply an eventual extension of the Harvard X, Y system including negative values.

(D) RADIO WORK

The work of the Sydney radio astronomers is yielding results of the greatest importance to our understanding of the two Magellanic Clouds. So far, surveys of the Clouds have been done for the 21-cm line by Kerr, Hindman and Robinson[8] and by Kerr and de Vaucouleurs[9, 10]; at 3·5 m in the continuum by Mills[11], and one has been done at 15 m by Shain[12]. It is planned to repeat the 21-cm survey with increased resolution in the near future.

The 21-cm observations have shown that there is a large envelope of neutral hydrogen around each Cloud; the rotation of each system has been demonstrated, confirming the view that the Clouds are flattened and tilted systems; but as previously mentioned improved values of the tilt are required. Further, from the 21-cm observations the masses and mass distribution have been estimated from the rotational characteristics and the gas is found to be probably the least flattened component of each system; the rotational velocity appears to vary with distance from the equatorial plane according to both optical and 21-cm observations.

The main 21-cm problems for the future are as follows:

(a) Study of the inner parts with higher angular resolution, to trace structural details;

(b) Study of the outer parts with higher sensitivity, including a search for a possible link between the Clouds and between the Clouds and the Galaxy;

(c) More detailed comparison of radio and optical velocities;

(d) Study of the variation of velocity dispersion across each system, where it may be simpler to separate such variation from other effects than it is in the Galaxy;

(e) Extension of the theory for the case of a system where the energy is fairly equally split between random and rotational motion.

In continuum observations, one difficulty is in the separation of the Clouds from the galactic foreground. In this case, observations of the Clouds may help to solve the problem of the origin of the radio continuum in the Galaxy, through comparing the distribution of the radiation over the Clouds with the distributions of various known classes of objects.

According to Mills[11] at 3·5 m, there is little or no sign of a corona round the Clouds, and the systems are underluminous in a radio sense, by comparison with galaxies of other types. The ratio of radio to optical luminosity is about the same for the two Clouds.

De Vaucouleurs[13] has made a comparison between various distributions. He finds that the distribution of the 3·5-m radiation resembles those of the interstellar gas and the number of bright stars rather than that of the luminosity; according to Mills[14] it thus appears to be related to population I.

A portion of the LMC has so far been studied at 15 m. by Shain[12]. The most striking feature is the absorption dip in the vicinity of 30 Doradus.

References

[1] Buscombe, W., Gascoigne, S. C. B. and de Vaucouleurs, G. *Austr. J. Sci.* **17**, 3, 1954.
[2] Henize, K. G. *Astroph. J. Suppl.* no. 22, 1956.
[3] Lindsay, E. M. *Mon. Not. R. Astr. Soc.* **115**, 248, 1955 and **116**, 649, 1956.
[4] Feast, M. W., Thackeray, A. D. and Wesselink, A. J. *Observatory*, **75**, 216, 1955.
[5] Feast, M. W. *Mon. Not. R. Astr. Soc.* **116**, 583, 1956.
[6] Feast, M. W. and Thackeray, A. D. *Mon. Not. R. Astr. Soc.* **116**, 587, 1956.
[7] Koelbloed, D. *Observatory*, **76**, 191, 1956.
[8] Kerr, F. J., Hindman, J. V. and Robinson, B. J. *Austr. J. Phys.* **7**, 297, 1954.
[9] Kerr, F. J. and de Vaucouleurs, G. *Austr. J. Phys.* **8**, 508, 1955.
[10] Kerr, F. J. and de Vaucouleurs, G. *Austr. J. Phys.* **9**, 90, 1956.
[11] Mills, B. Y. *Austr. J. Phys.* **8**, 368, 1955.
[12] Shain, C. A. (unpublished).
[13] Vaucouleurs, G. de. I.A.U. Symposium No. 4, 244, 1957.
[14] Mills, B. Y. 'Radio Frequency Radiation from External Galaxies', in *Handbuch der Physik* (in press).

IV. SUMMARY OF DISCUSSIONS AND DESIDERATA

The meeting was concluded by a summary presented by J. H. Oort and subsequent discussion of the principal desiderata for future work. The following paragraphs include the principal topics of this summary and the desiderata.

Galactic clusters and O associations

It has become more and more evident how extremely important the study of galactic clusters and O associations is for the investigation of galactic structure. This is so for three reasons:

(a) For locating spiral structure independent of the velocity. It would be hard to overemphasize this point. We can hope to determine the scale of the Galactic System from them, as well as spiral structure.

(b) Most important of all: they can give information on systematic deviations from circular motion.

(c) They lend themselves excellently for a determination of *age*, for a determination of change of stellar composition with time and for the derivation of semi-empirical paths of evolution.

The study of clusters may eventually teach us something about differences in distribution and motions of objects formed at various epochs in the development of the Galactic System. What we need is:

(1) To discover more distant clusters.

(2) To find older clusters.

(3) To obtain colour-magnitude diagrams of good accuracy for all clusters. This involves the problem of identifying members.

(4) Radial velocities.

In connexion with the latter item we refer to the report by Weaver on the valuable material on radial velocities of stars in galactic clusters obtained by Trumpler, which is now being prepared for publication.

At the instigation of Drs Heckmann and Haffner the feasibility of obtaining in a number of regions around the sky standard magnitudes and colours to be used for calibrating the photometric data for individual clusters was investigated during the meeting by a small committee. Its proposal, presented by Dr Walker, is given in the Appendix.

86

Cepheids

Kukarkin has stressed again the desirability of searches for more distant δ Cephei variables. There can be no doubt that this offers one of the most important ways of investigating galactic structure and dynamics. This has been borne out again by the report given by Oosterhoff on the colour-excesses of southern cepheids and the space distribution derived with the aid of these colours.

The amount of work involved in new surveys will be very large, and co-ordination is therefore important. It is evidently essential that accurate colour-excesses and radial velocities be determined for all δ Cephei variables that will be found.

Double and multiple stars

Kulikovsky has proposed that more extensive investigations be made of trapezium-type groups, as well as of wide double and multiple stars in general.

In connexion with the general topic of this conference mention should be made of the fact that a co-ordinating commission for T-associations was formed during the Burakan symposium on Non-stable Stars; members are Haro, Herbig and Kholopov.

Spiral structure

Since the Groningen conference a great mass of new information on the distribution of neutral hydrogen has become available. The observations concerning the southern parts of the Milky Way are now being reduced. A new development which appears particularly intriguing is the discovery of a spiral arm in the nuclear region which moves away from the centre, at a velocity of 53 km/sec. Other structures, closer to the centre, show still larger deviations from circular motion.

The report of the Groningen conference stressed the importance of studying possible differences between the distributions of A stars, K giants and the neutral hydrogen. Ramberg has reported on his work on A stars and K giants. Though indicating some relation with the nearby spiral structure of the gas the investigations do not yet show how closely the spiral structure of the gas and stars coincide. The investigations should be extended to larger distances and to more regions before they can give a final answer to this question.

A small working committee was appointed whose task it will be to investigate, how work on the density distribution for different components of the galactic population, in both low and high latitudes, can best be promoted. Its members are: A. Blaauw, W. W. Morgan, J. Ramberg, N. G. Roman, and A. Sandage.

Optical work on interstellar lines

The meeting endorsed a proposal by Kerr and Westerhout, urging the extension of optical work on the velocity distribution of interstellar clouds along the lines of recent work by Münch. Such work is of great significance for the interpretation of 21-cm observations.

H II regions

Kerr and Westerhout also urged extension of observations of H II regions, in view of the interpretation of continuous radio emission at short wave-lengths. More detailed radio surveys at wave-lengths shorter than 10 cm should supplement the available data. In this connexion, also, infra-red surveys are much wanted.

Special regions

Parenago has made a proposal to study a number of special regions (see p. 61). Undoubtedly it is of considerable value to concentrate efforts on a few special regions, and those selected are undoubtedly important ones. It was therefore proposed that the conference agrees in principle to the plan and advocates that support be given to it as far as possible. It was stressed at the same time that attention should be concentrated on individual data of very great accuracy rather than on rough data for great numbers of stars. It is still uncertain how much value the determination of proper motions will have in connexion with the study of the space structure of these regions.

Accurate spectrophotometry

The possibility of accurate spectrophotometric determinations of spectral types, luminosities and possibly a third parameter has been discussed by Chalonge, Strömgren and Miss Roman. The subject had been much discussed at our previous symposium; since then a large amount of data has been collected and the great advances that may be obtained through this type of accurate measurement have been amply demonstrated. Here indeed is an enormously rich mine for observational programmes of the greatest importance for galactic research.

Miss Roman's investigations of high-velocity stars have indicated that the ultra-violet excess is well correlated with average velocity, and may therefore be an important indicator of age for the types of stars she studied.

O. C. Wilson has recently shown that the width of the emission core of the K line in G, K and M stars is strongly correlated with absolute magnitude, while being independent of spectral type. This might prove to be the most accurate method for determining absolute magnitudes. But high dispersions are required. Co-ordination with other accurate measures of luminosity is important.

Trigonometric parallaxes

It was proposed that parallax observers would be asked to determine accurate trigonometric parallaxes of about two dozen stars with very weak lines, a list of which will be made available by Miss Roman.

Determination of K_z

This still leaves much to be desired. The best prospects would seem to be offered by accurate objective-prism classifications of K stars down to about 14·0 photographic magnitude in the galactic polar caps, preferably in Kapteyn's Selected Areas where photometric and proper-motion data are already available. It would be desirable to make measurements that would separate giants from sub-giants and dwarfs. Knowledge of radial velocities down to the same magnitude limit would be of the greatest importance.

Radial velocities

Progress with the objective-prism method, as reported by Fehrenbach, is extremely promising. Similar work will be taken up at Herstmonceux. Co-operation for obtaining a sufficient number of faint standard velocities is desirable.

Nucleus and disk

The desirability of paying more attention to long-period variables for investigating the older disk population was stressed by Kukarkin and strongly supported by the meeting.

Halo

The search for RR Lyrae variables in a number of regions in moderate galactic latitudes which was advocated at the Groningen conference has meanwhile been undertaken, and has been reported at this meeting. Complete sets of plates down to a faint magnitude limit have been obtained with the 48-inch Schmidt of Mt Palomar and are now being searched for variables by Dr Plaut at Groningen. An essential desideratum still to be provided is colour excesses for all the variables found.

Morgan has reported on the striking difference in the spectra of various globular clusters. These indicate that there is a class of clusters which are relatively rich in heavy elements and which is probably concentrated towards the galactic disk. However, there are likewise considerable differences in composition among clusters having large distances from the galactic plane.

Further investigations of spectra, as well as colours and magnitudes of individual stars in globular clusters, are highly desirable.

Magellanic Clouds

Thackeray reported on some of the work on the Magellanic Clouds that is now being done or that would seem of importance to be done in the nearest future. This latter comprises a.o. objective-prism surveys to 13·5 photographic magnitude, objective-prism radial velocities to eliminate foreground objects, a search for bright variables, further photometry of known variables and a general search for new ones, in particular also for those of long period. An atlas of charts should be made, covering a larger field than the Harvard charts.

APPENDIX

REPORT OF THE COMMITTEE ON PHOTOMETRIC STANDARDS

(PREPARED BY M. F. WALKER, CHAIRMAN)

In view of the high precision which is now both possible and desirable in photometric observations, it is extremely important for the co-ordination of galactic research that all observers adopt a uniform photometric system. It is likewise important for reasons which have already been discussed at the Dublin meeting of the I.A.U. [1] that such a system should include a measurement of the ultra-violet, as well as the usual yellow and blue regions. The zero-point and scale of the system should be set up photo-electrically and standards on the system should be made available that would be suitable for the observation of faint objects of all kinds and in all parts of the sky.

To implement these needs, the committee makes the following recommendations:

I. *Photometric System.* It is recommended that the U-B-V system of Johnson and Morgan [2] be adopted. Improvements in the system could undoubtedly be made, both in the choice of the wave-length regions and in the filters used to define these regions. However, the greatest need is for *a* standard system *now*, and since the U-B-V system now has behind it a considerable weight of observations of a wide variety of objects, it is felt that this system should, at least for the present, be adopted in its present form.

In order to reproduce the U-B-V system, the following filters are recommended:

(*a*) For photo-electric observations with a 1 P 21 photo-multiplier or other photo-electric cell of similar spectral response, Johnson [3] has recommended the following filters: (1) V, Corning 3384 (Standard optical thickness) or 2 mm Schott GG 11. (2) B, Corning 5030 (Standard optical thickness) plus 2 mm Schott GG 13, or 1 mm Schott BG 12 plus 2 mm Schott GG 13. (3) U, Corning 9863 (Standard optical thickness) or 2 mm Schott UG 2.

(*b*) For photographic observations, the following plate and filter combinations have been found by Johnson and Sandage [4] to give no color equation to the U-B-V system: (1) V, Kodak 103aD plus 2 mm Schott GG 11. (2) B, Kodak 103aO plus 2 mm Schott GG 13. (3) U, Kodak 103aO plus 2 mm Schott UG 2.

Other filters—and plates—closely similar to these can also be found that will give linear transformations to the U-B-V system. However, whatever filters are used, the observer should in every case determine the conversion equation between his particular instrumental system and the U-B-V system, as discussed under Technique, below.

II. *Technique of Observation.* In order to avoid the problems of scale errors and to minimize the errors in the transfer of the zero-point of the system from one part of the sky to another, it is recommended that the use of *photographic* transfers be discontinued except where magnitudes of relatively low accuracy for statistical purposes *only* are required. Where a large number of faint stars are to be observed in a small area of the sky, as in the case of clusters, photo-electric transfers should be made to the area, and a photo-electric sequence set up within the area. Photographic photometry should be used only as an interpolation method for determining the magnitudes of the other stars in the field from those of the photo-electric standards. If the particular observatory does not have the facilities for setting up faint photo-electric standards, it is recommended that a qualified observer be sent to some other institution having such facilities to make the measurements. Representatives of the Haute-Provence and Mt Wilson observatories have indicated that guest investigators could arrange to make such observations at these institutions.

As indicated above, each observer should determine the transformation equations from his particular instrumental system to the U-B-V system, both for his photo-electric and photographic equipment. To determine the photo-electric conversion equation, about twenty stars from the lists of U-B-V standards [2, 3, 5] should be observed. These should cover a large range of colors and luminosity classes. One defect of the U-B-V system is that the standards are nearly all very bright, so that it is difficult to observe them with a large telescope and a photo-electric cell suitable for the observation of extremely faint stars. However, for calibration purposes, they can be observed by the use of objective screens (but not diaphragms, which could image the light on a different part of the cathode having a different sensitivity and color response from the average for the cathode as a whole) or by operating the cell at a reduced voltage. An objective screen will usually be found to have an absorption that depends slightly on wavelength, and this must be allowed for. Also, some types of photo-electric cells may show a change in color response with applied voltage, although this is not usually the case.

To determine the conversion equations, one should observe a field containing stars whose U-B-V magnitudes have already been determined

In order to discriminate between magnitude and color equations, these fields should not only contain a sequence of stars of different magnitudes, but stars of comparable magnitude and different color as well. At the present time, such fields include M 11 [6], M 67 [4], and NGC 7789 [7]. Once the photo-electric conversion equations have been determined, observations of a red and a blue standard on each night will give the zero-point of the conversion equation for that particular night. The photo-electric and photographic conversion equations should be carefully rechecked from time to time as there is the possibility that they may change with season, changes in the emulsion number, or the condition of the optical surfaces of the telescope.

III. *Standards*. In view of the foregoing recommendations, it is evident that what we require at the present time are photo-electric standards distributed over the entire sky. As indicated above, the standards on which the U-B-V system is based are not well suited for use with large telescopes since they are too bright to be observed with the same cell voltage and aperture used for fainter stars. Consequently, standards fainter than about 7th magnitude should be made available. Also, because of practical limitations of some telescopes, it is desirable to have standards located both north and south of the zenith. As an initial program, it is therefore recommended that the following standards be set up: A red and a blue standard of about 8th magnitude and another red and blue pair of about 10th magnitude should be set up in each of the following areas: (*a*) Decl. = $+45°$, Δ R.A. = 4^h. (*b*) Decl. = $0°$, Δ R.A. = 3^h. (*c*) Decl. = $-45°$ (stars in the E regions). Standards in areas (*a*) and (*b*), above, will be set up by Hardie and Walker. The standards in (*c*) will be provided by Haffner in South Africa, who will make use of the extensive work already done in the E regions at the Cape Observatory. Haffner will also observe the stars in areas (*b*), to complete the tie-in between the northern and southern hemispheres.

In addition to the above areas, the red and blue stars, of about 8th to 12th magnitude, used by Baum as his fundamental standards in each of the nine selected areas 51, 54, 57, 61, 68, 71, 89, 94 and 107 in which he is setting up faint two-color photo-electric sequences, will be observed by Hardie and Walker.

References

[1] *Trans. I.A.U.* **9**, 338, *et seq.*, 1955.
[2] Johnson, H. L. and Morgan, W. W. *Astroph. J.* **117**, 313, 1953.
[3] Johnson, H. L. *Ann. d'Astroph.* **18**, 292, 1955.
[4] Johnson, H. L. and Sandage, A. R. *Astroph. J.* **121**, 616, 1955.
[5] Johnson, H. L. and Harris, D. L. *Astroph. J.* **120**, 196, 1954.
[6] Johnson, H. L. and Sandage, A. R. *Astroph. J.* **124**, 81, 1956.
[7] Burbidge, E. M. and Sandage, A. R. In the press.

Printed in the United States
By Bookmasters